Dolphin Mysteries

KATHLEEN M. DUDZINSKI, PH.D.   TONI FROHOFF, PH.D.

DOLPHIN MYSTERIES

Unlocking the Secrets of Communication

ILLUSTRATIONS BY JOHN NORTON

PHOTOGRAPHY BY JOHN ANDERSON

YALE UNIVERSITY PRESS • NEW HAVEN AND LONDON

Published with assistance from the foundation established in memory of Philip Hamilton McMillan of the Class of 1894, Yale College.

"Baby Beluga," words by Raffi and D. Pike; music by Raffi; © 1980 Homeland Publishing (SOCAN), a division of Troubadour Music, Inc. All rights reserved. Used by permission.

"Whales Weep Not!" by D. H. Lawrence, from *The Complete Poems of D. H. Lawrence*, by D. H. Lawrence, edited by V. de Sola Pinto and F. W. Roberts, copyright © 1964, 1971 by Angelo Ravagli and C. M. Weekley, Executors of the Estate of Frieda Lawrence Ravagli. Used by permission of Viking Penguin, a division of Penguin Group (USA) Inc., and Pollinger Limited (UK).

Bizarro cartoon copyright © 2008 by Dan Piraro.

Designed by Nancy Ovedovitz and set in Janson Oldstyle by Duke & Company, Devon, Pennsylvania. Printed in the United States of America.

Library of Congress Cataloging-in-Publication Data
Dudzinski, Kathleen.
Dolphin mysteries : unlocking the secrets of communication / Kathleen M. Dudzinski and Toni Frohoff.
    p.   cm.
Includes bibliographical references and index.
ISBN 978-0-300-12112-4 (clothbound : alk. paper)
1. Dolphins—Behavior. 2. Animal communication. I. Frohoff, Toni, 1963– II. Title.
QL737.C432D836 2008

599.53′159—dc22   2008017678

A catalogue record for this book is available from the British Library.

This paper meets the requirements of ANSI/NISO Z39.48-1992 (Permanence of Paper).

10 9 8 7 6 5 4 3 2 1

KMD *For John, who often understands my signals even when I do not*
      For Umi, who has taught me that actions often speak louder than words

TF   *For my loved ones, who have been like a pod to me, and for the dolphins*
      who continue to guide my way without saying a word

Contents

Color plates follow p. 100.

Foreword

Dolphin Mysteries is a wonderful and very welcome book about a "poster child" of sentient beings. But it's really many books in one. Its scope is truly enormous, and Kathleen Dudzinski and Toni Frohoff are to be congratulated for getting so much important and timely information between two reasonably spaced covers. What sets this book apart from many others that deal with dolphins and their relatives is that not only is it written by two experts—two passionate women who have devoted much of their waking (and perhaps their dream) lives to dolphins—but it presents a diverse array of information that surely will be educational to a broad audience, academic and lay people alike. *Dolphin Mysteries* will appeal to anyone who wants to know more about these amazing animals and will inspire those who want to improve the lives of captive dolphins and their more fortunate relatives who live free, or almost free, of human intrusions. It is scientifically rigorous, is easy to read, and has plenty of heart. What a wonderful combination.

As I read this book, I often thought of Thomas Berry's claim that a group of individuals really is a communion of subjects, not merely a collection of objects. This surely is true for dolphins and many other animals who must be able to live in harmony with other sentient beings, including those who persecute them. Kathleen and Toni show clearly that dolphins are amazing individuals. Most descriptions of dolphins and other cetaceans portray them as highly intelligent, sentient animals with remarkable social and cognitive skills. They're highly emotional, playful, seem to empathize with one another, have a sense of self, and mourn the death of other dolphins. We now know that individuals of many species use tools, have culture, are conscious and have a sense of self, can reason, can draw, can self-medicate, and show very complex patterns of communication that rival what we call "language."

As you read this fine book and learn more about the world of dolphin communication and cognition, you'll see that dolphins are more remarkable and mysterious than you could have imagined. By going beneath the surface, experts Kathleen and Toni open up doors of perception and give us a glimpse into the world of dolphins from both beneath the water's surface and behind the dolphin's smile. In the spirit of classical ethologists such as Konrad Lorenz and Niko Tinbergen they ask, "What is it like to be a dolphin?" and provide up-to-date information about numerous aspects of dolphin natural history, behavior, behavioral ecology, sensory ecology, intelligence, conservation, and anatomy and physiology.

A few years ago I published an essay in a book titled *Intimate Nature* that was coedited by Toni and Brenda Peterson. My essay was called "Troubling *Tursiops*," an obvious play on the genus of bottle-nosed dolphins, *Tursiops truncatus*. The title has two meanings. The first twist on the word "troubling" concerns how humans trouble *Tursiops*, how we intrude into their worlds, how we bother them intentionally and unintentionally. The second twist on this word concerns the troubling and complex issues that arise when we ponder what sort of intrusions, if any, are permissible. I began my essay by describing how I was once standing in line at a grocery store and overheard a girl tell her friend that she'd just gone swimming with dolphins when on holiday in Hawaii. She had a great time, but there was a slight pause when her friend asked her about what the dolphins might have felt about all of this. She asked, did they enjoy her touching them or riding on their back? Did they really like being bothered? I was pleased to see questions about ethics being raised by a youngster.

As a scientist, I believe that it is in the best traditions of science to ask questions about ethics. It's not antiscience to question what we do when we study other animals. Ethics can enrich our views of other animals in their own worlds and in our different worlds, and it can help us to see that variations among animals are worthy of respect, admiration, and appreciation. The study of ethics can also broaden the range of possible ways in which we interact with other animals without compromising their lives. Ethical discussion can help us to see alternatives to past actions that have disrespected other animals and, in the end, have served neither us nor other animals well. In this way, the study of ethics is enriching to other animals and to ourselves in that we may come to consider new possibilities for how we interact with beings with whom we share our planet. If we think ethical considerations are stifling and create unnecessary hurdles over which we must blindly jump in order to get done what we want to get done, then we'll lose rich opportunities to learn more about other animals as well as ourselves.

Cetaceans are closely linked to the wholeness of many ecosystems, and how they fare is tightly associated with how communities and ecosystems fare. Dolphins are also closely linked to our own spirituality. By paying close attention to what we do to them and why we do what we do where and when we do it, we can help maintain the health of individuals, species, populations, ecosystems—and ourselves. There's no substitute for respecting animals, treating them with heartfelt compassion, and loving them for who they are in this magnificent and awe-inspiring world. Could anyone reasonably argue that a world with less cruelty and destruction and more compassion and love wouldn't be a better place in which to live and to raise children? I don't think so.

Public education is critical. So, we need to be sure that our knowledge about other species gets out to a broad audience. We really need books like Kathleen and Toni's to give people the information they need to be knowledgeable and proactive advocates for making the life of dolphins better worldwide. To disseminate information about what is called the "human dimension," administrators of zoos, wildlife theme parks, aquariums, and areas where animals roam freely need to inform visitors of how their behavior influences the behavior of the animals they want to see. Tourism companies, nature clubs and societies, and schools can do the same. By treading lightly, we humans can enjoy the company of other animals without making them pay for our interest in their fascinating lives.

It's truly a privilege to study animals and to share their worlds and lives with them. As we learn more about how we influence other animals we will be able to adopt proactive, rather than reactive, strategies. Part of learning entails changing our practices and asking, "Would we do what we did again?" and "Have we learned something that can make other animals' lives better?"

Our curiosity about other animals need not harm them. We *can* live in peace with other animals and easily learn about who they are in their and our worlds. Our old brains crave harmonious and peaceful connections with other nature, and such encounters are good for all of us. When animals lose, we all lose. When we have friendly interactions with other animals, everyone wins, and with hard work we can work together to make the world more compassionate and peaceful, a point I stress in my book *The Emotional Lives of Animals*. It's pretty amazing what a group of right-minded people can do.

Many thanks to Kathleen and Toni for taking the time out of their busy family and professional lives to write a book for all of us: one that will undoubtedly make a difference in how dolphins—and, let's hope, other animals—are perceived. Even when we think we're doing the best we can, I like to say that isn't enough—that "good welfare" isn't "good enough." We can always do better, and we can always do more in our interactions with animals. That's my guiding light, my bumper sticker. And now, at last, there is a book that also serves as my guide. I hope it will guide and inspire numerous people worldwide.

Marc Bekoff, Ph.D.
Fellow of the Animal Behavior Society
Boulder, Colorado

Preface

Are you ready to take the plunge? Dolphins come in a wide variety of packages, ranging from the tiny Hector's dolphin, at about 4.5 feet (1.5 m) long, to the almost 30-foot (9-m) killer whale. They are found in environments that range from vast offshore ocean habitats to rivers to coastlines to pools and pens. Dolphins are highly gregarious, social animals with both consistent and fluid societal relationships. The nature of these relationships, as manifested in group size, cohesion, and composition, seems related to their habitat and geography. Some aspects of a dolphin's life are amazingly similar to that of other social animals—aquatic and terrestrial, wild and domesticated. Dolphins are, however, truly their "own animal" with respect to the many unique ways in which they interact with their environment, other dolphins, and even other species. And the more we study individual dolphins, the more they express their uniqueness to us through what we have come to know as "dolphinalities."

Our interests in dolphin communication originated from different experiences that eventually converged into a similar passion for "eavesdropping" on dolphin social behavior. Our individual paths transported each of us toward the underwater study of dolphins with an emphasis on their communication and interactions both with one another and with people. We share a vision of diving beneath the surface with dolphins to understand the intricacies of their society where they reside. Plunging beneath the waves lets us go beneath the surface of the dolphin smile; our studies of their behavior and communication give us a window into the mind, heart, and inner life of the dolphin.

Although dolphins spend about 99 percent of their lives beneath the ocean's surface, there is no comprehensive text presenting and discussing dolphin behavior primarily from the underwater perspective. Most of the literature in peer-reviewed scientific articles and popular books addresses dolphin behavior based on data gathered solely from above the sea surface, from boats, or from land. Through our work we have always, whenever possible, focused on dolphin social lives, especially their behavior and communication, through an underwater lens. With the advent of new technology and the assistance of dolphins willing to put up with our "spy games," we have been increasingly able to observe and document the behavior and sounds of dolphins from this underwater vantage point.

Communication—the exchange of information—binds all animal societies together. Dolphins live within structurally coordinated social assemblies where communication among individuals and groups is important for the maintenance of social life. The behavior of dolphin societies reflects a dynamic interplay of altruism, aggression, sexual interactions, cooperative hunting, exploratory behavior, play, flight, and predator avoidance. One way to learn about their communication is to "eavesdrop" on their social activities.

But how does one eavesdrop on a dolphin? Investigations of dolphin-to-dolphin communication give us clues to the etiquette and rules that govern dolphin social life. We find that although studies of group behavior conducted from above the surface are certainly valuable, they often exclude the subtle, myriad, and richly complex actions exhibited by individual dolphins, most of which can be witnessed only underwater. Here we can explore the visual and vocal signals that dolphins use to share information with one another. Opportunities that lend themselves to this eavesdropping include captive dolphin facilities and dolphins in the wild who have habituated to (or at least tolerate) the presence of swimmers and researchers. Unique opportunities for close, detailed, intimate, and lengthy observations of previously unknown or poorly understood dolphin behaviors are now possible. For us, eavesdropping represents a tool for gaining access to the intimate social lives and behavior of another species.

The study of communication is the perfect vehicle through which to navigate a better understanding of cognitive processes and emotion in dolphins. In fact, examining the signals that dolphins use to express themselves is necessary in order to see beyond their smile. Although emotion and consciousness are difficult qualities to measure and to confirm empirically, compelling evidence is mounting that supports the existence of rich emotional lives and a world of inner conscious experience by many nonhuman animal species. In the past several years, science articles in major international newspapers have appeared with such titles as "Animals Enjoy a Good Laugh too, Scientists Say," and "The Secret Life of Moody Cows." It comes as no surprise that scientists are studying the thoughts and feelings of dolphins.

The new millennium has brought discoveries about dolphin cognition, emotion, and consciousness, creating a bridge to what is known about dolphin behavior with an emphasis on communication. New data on dolphin intelligence allow us to obtain a broader picture of the inner lives of dolphins. We see that dolphins exhibit sophisticated characteristics previously attributed only to humans and possibly to other higher primates. We know that dolphins understand syntax, semantics, and word order and are capable of mirror-self-recognition, comprehension of pointing gestures, and understanding reference to body parts. They can identify the same abstract object using vision or echolocation. There is evidence that dolphins have culture. For instance, killer whales have vocal dialects that are distinct to each family group and are passed down through generations; some bottlenose dolphins use sponges as tools when foraging along the sea floor.

The study of dolphin interactions with people provides a rich and invaluable lens through which we can gain a unique perspective on dolphin communication, behavior, sociality, intelligence, and emotion. Dolphins are rare among wild animals in that it is not uncommon for them to approach humans or initiate "sociable" interactions with people. Given the fascinating and often misunderstood nature of this topic, it is particularly challenging to limit our exploration of dolphin-human interactions and interspecies communication. Yet by applying the tools of science to our observations of dolphin-human interactions, we can unravel some of the subtleties of dolphin communication and reveal what may be some of the most intriguing aspects of dolphin life. The dolphins who initiate friendly contact with people and occasionally form close bonds with humans over time, are most often solitary individuals who are rarely, if ever, observed in the company of other dolphins.

We call these individuals "solitary, sociable" or "lone, sociable" dolphins. Studies documenting these interactions are shedding new light on historical anecdotes of dolphin-human relationships that have accumulated since ancient times. Many scientists have ignored the subject of "interspecies communication" largely because of the public glamorization of this topic in the 1960s and 1970s, when this term was coined. Over the past two decades, however, we have collectively revisited this subject using established ethological techniques and a renewed scientific rigor.

An overview of historical and modern-day research on dolphin-human interactions presents a comparative view in reference to what we know about dolphin-to-dolphin communication. For instance, we discuss why dolphins interacting with human swimmers direct species-specific behaviors at people but may also modify their signals to "accommodate" humans. We share examples in which dolphin-human interactions are more indicative of *mis*-communication than communication. We explore the mythology surrounding dolphins and see that, surprisingly, much of it may be related more to fact than to fiction. Simultaneously, we explore a different side of dolphins that contradicts much of what we know about them from fable and pop culture. For instance, dolphins are known to assist humans stranded in the oceans, conceivably driven by social curiosity to save the life of an animal not their own kind. In contrast, dolphins are also known to act aggressively toward humans, even perhaps to injure or kill. This complexity and the individuality that dolphins exhibit are what make the facts about them even more interesting than the myths that surround them.

We conclude our voyage into the dolphins' world with a discussion about how the study of communication can lead to improved conservation practices. Many dolphins face serious wildlife conservation challenges, and some species are on the verge of extinction. Dolphins are intentionally killed, legally and illegally, in many parts of the world. In June 2007, several bottlenose dolphin bodies washed ashore in southern California; they had been shot illegally. Investigators suggested that the dolphins may have been target practice for boaters reveling over a holiday weekend. Other formidable threats to dolphins include reduction of their food supply because of overfishing and bycatch, climate change, and habitat degradation in the form of harassment by and collision with seagoing vessels, incidental *weird farms?*

mortality during fishing operations, anthropogenic noise emissions (from boats, oil exploration, and military activities), and chemical and debris pollution. A plethora of disturbing environmental information is being reported in the popular media and in scientific journals. Yet we find encouragement in how research on dolphins, when implemented into effective policy and management, can empower us to mitigate not only the threats faced by these incredible mammals but also the environmental threats shared by all of us with them.

In writing this book, we surprised even ourselves in seeing how integral the study of dolphin communication is to the recognized structures of wildlife research, welfare, and conservation. Ethical and scientific implications for dolphins in captivity and conservation of dolphins in the wild are presented with respect to life in a world in which humans are increasingly affecting the marine environment. What, we ask, is our ethical responsibility as scientists to our fellow resident animals on earth?

The better we understand the signals used to coordinate communication and social activity among individuals within any population, no matter the species, the better we will understand the links and separations that underlie the evolution of social life and survival strategies. More specifically, increasing our knowledge of dolphin social ecology and communication heightens our understanding of the dynamics of their society and how we, as human interlopers, affect dolphins' development individually and as a community. Understanding our effects helps us better manage our actions and behavior with respect to these ocean dwellers in an increasingly fragile habitat—potentially leading to better conservation and protection practices for dolphins and all marine life.

One of the most intriguing questions we have received over the years is "Will we ever share a language with dolphins?" We ask readers to consider a slightly different question: "Do we even need a concise mutual language, or can we achieve a true, viable understanding of dolphins by eavesdropping on and learning from *their* communication?" As a souvenir from this undersea exploration, we hope to leave you with an enhanced sense of fascination and respect for the countless sights and sounds of the dolphin and a platform from which to jump further into the world of these amazingly expressive creatures of the deep.

Acknowledgments

Before entering the world of dolphins, we must thank several individuals for their support, wise counsel, and inspiration on our respective career, indeed life, paths. At the start of both our academic careers, almost twenty years ago, two people guided us in establishing the foundation for our scientific thought, reasoning, and application. Bernd Würsig and Jane Packard, our primary graduate advisers in the Department of Wildlife and Fisheries Science at Texas A&M University, shared with us their remarkable skills and passions. They are also the likely reason we met and set the stage for both our solo and joint ventures and adventures over the years. In graduate school, Marc Bekoff inspired us with his innovative books and articles—required reading! Marc's input and advice over the past fifteen years has helped us shape this book, and we are delighted that he wrote the foreword.

Friends and colleagues provided constructive, thoughtful, and thought-provoking critiques of early drafts, making this book a better read and helping us see our research and passions with fresh eyes. Thank you, Leigh Calvez, Suzanne Chisolm, Sandra Dudzinski, Dagmar Fertl, Stan Kuczaj, and Michael Parfit. Justin Gregg contributed to and reviewed our prose for scientific accuracy especially in the areas of cognition, linguistics, emotion, and related studies. Jillian Kasow was ruthless in her attention to grammar and punctuation and helped us to liven our technical writing. Both Justin and Jillian were essential to the completion of this book.

As our careers developed and our niches settled, colleagues inspired us through lively debate, entertaining discussion, and thoughtful discourse in the field, in the lab, at a conference, or just hanging out. Thank you to the marine mammal community collectively for the body of knowledge and information that grows with each passing year.

Two artists have enhanced tremendously our ability to guide readers on this underwater journey. John Anderson's brilliant photography and underwater insights at times represent the dolphins better than our descriptions. If a photograph is worth a thousand words, then our book is double the agreed-upon length! John Norton's masterful illustrations bring greater life to the text. A joy to work with, John has been generous with his time and his exceptional talent.

We thank Dan Piraro for permission to include one of his wonderful Bizarro comics that has graced each of our offices for years. Cathy Kinsman of the Whale Stewardship Project and Will Anderson kindly contributed photographs of the belugas and gray whales they have studied and worked so hard to protect. Sally Antrobus and Emily Gustafson provided invaluable and expert editorial assistance.

Neither words, whistles, nor clicks can express how grateful we are to have worked with Jean Thomson Black, our editor at Yale University Press, who expertly embraced, steered, and encouraged us as we wrote, edited, and rewrote some more. We thank Laura Jones Dooley for copyediting our manuscript in a way that polished our written words without taming them. We appreciate the vision, support, and enthusiasm that the staff of Yale University Press has given to our book.

Our research over the years has been supported by a variety of sources. We have both received financial support from Ocean Society Expeditions, the National Geographic Society, and Texas A&M University. Kathleen's research has been supported by the National Science Foundation, the Lerner-Gray Foundation, the International Women's Fishing Association, Seaspace Association, Cetacean Society International, the Japan Society for the Promotion of Science, ICERC Japan, the Deramus Foundation, Mystic Aquarium, and the Dolphin Communication Project. Toni's research has been supported by TerraMar Research, the Summerlee Foundation, the Whale and Dolphin Conservation Society, the Whale Stewardship Program, Patagonia, Baja Expeditions, Palm, Inc., Save the Elephants, the Wild Dolphin Foundation in Oahu, Henderson Dive Wear, Isabel Jordan of

Yelapa, Dolphin Expeditions, and various government and nongovernment agencies around the world.

Like the dolphins beneath the waves, many people have assisted us along this adventure, and we are that much better for having traveled with them. The animals and humans we have each observed for more than twenty years continue to give us hours of enjoyment and purpose. Thank you.

My parents, Sandy and Pete, provided my sisters and me with an independent streak and the drive to learn and excel at anything. They gave me the foundation a good scientist needs. My sisters, Rebecca and Stephanie, continue to inspire me with their creativity, loyalty, compassion, and humor.

My work has taken me to many countries, and I have an extended family of friends. Tomoko and Takeshi Ootake and the Yamamotos showed me the true meaning of cultural exchange and welcomed me into their homes and hearts both when I lived in Japan and now. Motoi Yoshioka, Kazunobu (Mampu) Kogi, and Tomoko Fujimaki helped me blend more detail into my life studying dolphins in Japan. Sunna Edberg gave me a roof, shared supper, and welcomed me into her dolphin world in Sweden. Teri and Eldon Bolton introduced me to Esteban, Paya, Rita, Gracie, and the rest of the gang (delphinid and human) at the Roatan Institute for Marine Sciences and continue to welcome me back. Annette Dempsey and Kelly and Robert Meister extend the same welcome to me at Dolphin Encounters, Nassau, The Bahamas. The list could go on for pages. I thank both those mentioned and those not for their continued support.

My students—graduate, undergraduate, and secondary school level—delight me with their dedication, enthusiasm, and ambition. In them I see wonderment and excitement about our natural world. Through them I have hope for our ocean planet.

I met John—my best friend, my husband, my sweetheart, my photographer—because of dolphins. Without him, my life would be drab and dull. I am grateful for him for reasons too numerous to list.

And last, but most certainly not least, Umi, the mighty sea beagle, is my sweetie pie. Her antics, smiles, howl-barks, grass temper tantrums, snubs, "rubbee" solicitations, affection, nose kanji, and snotty-ness keep me enthralled. Each day, I see behavior with a fresh perspective simply by watching Umi. Woof.

Kathleen M. Dudzinski, Ph.D.
Stonington, Connecticut

I acknowledge the many people who, like dolphins, have graciously lent their support and vision to me in various ways. In my earlier work Marc Bekoff, Denise Herzing, and Karen Pryor provided pivotal guidance and mentorship that continue to guide me. I have learned much from many others, including Cathy Kinsman, whom I have been blessed to continue to work and grow with, and Brenda Peterson, whose writing inspires me with its stunning and natural beauty. Naomi Rose, Sharon Young, and Erich Hoyt have taught me much about commitment to scientific accuracy and advocacy, and their contributions to this book are sincerely appreciated. I have been privileged to work with many dedicated people, including Cathy Williamson, Courtney Vail, and Mark Simmonds. I extend my gratitude to Sharon Negri and Leigh Calvez, my courageous, creative, and compassionate companions in wildlife adventures; to Bill Rossiter for his insights and generosity; to Christine Lamb for her artistic talent; and to Randi Curtis, Janette Wilson, Robin Lindsey, and Julie Dennis for their generous hearts and invaluable creativity and skill. I thank my coauthor, Kathleen Dudzinski, for inviting me to join her in the journey of writing this book and for the enthusiasm and brilliance she has shared with me over years of working together.

My life has been graced by the love of a constellation of immense hearts and wisdom from near and far. Among these have been Lilliana and Angela Zuyovich, Susan and David Scheirman, Jim Frohoff, Will Anderson, Jan Bailey, Mary Stowell, Michael Donais, Leah Lemieux, Joe Olson of Cetacean Research Technology, Stephen R. Goodman, Jim Dowling, Karl Straub, and my beloved Gudish family, sister Carol, brother Donnie, and our shared family, including little Lillian Rose. I cannot forget Sheppy, the best canine editorial assistant a writer could have.

When counting dolphins from above the surface, there are always some underwater—but neither should be overlooked. Similarly, I extend my appreciation to the many who have left an indelible impression on my life but whose names are simply not visible on the surface of these pages.

Toni Frohoff, Ph.D.
Puget Sound, Washington

Dolphin Mysteries

Introduction

I have always loved the ocean. We went to Cape Cod every summer when I was a child, and it was during these summers that my passion for learning and science was seeded. I remember long walks on the beach with my dad, scanning the ocean with binoculars and enjoying the waves and sandbars. I remember collecting shells with my mom. I treasure the nature hikes and the nooks and crannies of the dunes and seashore that my family "discovered." My love for animals also came from my family. The path that would lead to my career studying dolphin communication, however, did not become clear until I entered college. It is a path rooted in family history but forged with friends and colleagues.—Kathleen

When I first heard and saw dolphins underwater, I woke to a realm I had not known existed. This experience not only expanded my perception of the world but reshaped my life. My career as a scientist studying dolphins and being in their presence has revealed so much more than the data I sought to obtain. Dolphins are among my greatest teachers; they have guided me to a deeper understanding of what it means to be human in a world that is so much more than human. Through this interspecies exploration, dolphins continue to beckon and challenge me to deepen my perception as a scientist.—Toni

Dolphins entice us clumsy land-based humans with their aquatic grace, agility, beauty, and mystery, as well as what some have described as an unearthly intelligence and communication system. Over the years, a belief has been fostered that dolphins are directly comparable to people, with hints that they might have humanlike personalities. When we encounter them by boat or from land, we perceive them from our respectively different worlds. When we meet them in their world, by immersing ourselves in their aquatic habitat, two very different species connect on a common playing field.

We continue to learn, be inspired by, and remain in awe of these mammals we study collectively as species and those we have come to know as individuals. It is our privilege to take you with us on our journeys beneath the waves with these complex, perplexing, and perpetually fascinating animals. It is a way for us to share our passion, at times an absolutely consuming obsession, with our quest for a more detailed, comprehensive, and holistic understanding of dolphin communication and social behavior. Part of the beauty of this voyage is how, by studying dolphins in unique ways, we have obtained new perspectives with which to navigate the myriad avenues of conservation and dolphin protection. We invite you to join us on an adventure to become more aware of the intricacies of dolphins' social lives through personal accounts of our experiences and observations, as well as through our presentation of the most accurate information available about dolphin communication, cognition, culture, emotion, and intelligence. We present our stories in the first person and acknowledge that our use of anecdotes is aimed at illustrating the facts gleaned from our colleagues' research as well as our own. As the animal behavior expert Marc Bekoff has stated, "The value of a reliably reported anecdote should not be underestimated, even in science." Our data from almost forty years of collective research form the foundation for our words. To prepare for immersion into the world of dolphin research, we wish to provide a bit more background into the influences on our lives and into the sometimes controversial history swirling around studies of dolphin communication.

Kathleen

I found dolphins as a college student on a summer internship in Gloucester, Massachusetts, in 1987. I did not love dolphins from birth, though I readily admit to loving animals, the ocean, and school from a young age. Science was always my favorite subject: I was a nerd, a geek, from middle school forward.

Even now, I could list for you every science teacher from seventh grade to college, and I am not good at remembering names. Two secondary school teachers, Ms. Chamberlain and Mr. Faustman, stand out in my memory as exceptional. They made me think and work hard for every bit of knowledge I acquired. I credit my parents and these two teachers with laying the foundation I needed to become a scientist. They fostered in me a passion for learning and for questioning my observations.

Ms. Chamberlain was my eighth-grade science teacher. We made model houses that were powered or heated by alternative energy sources (she was ahead of her time!) and learned to make soaps from natural products. Tenth-grade biology was Mr. Faustman's class, and we had a blast. My most vivid memory of biology with him was the preparation of our spring "Anatomical Foods Night": each student had to create a dish that was all natural, related to an anatomy lesson, or traditional to our family heritage. Milk from two local farms, haggis, heart, liver, seaweed candy, soy noodles, and other items were feasted on by all who attended. Each dish came with a lesson rooted in biology, and Mr. Faustman provided a comical slide show depicting our efforts as the evening's entertainment.

The best compliment I have ever received was from my parents, Sandy and Pete Dudzinski. In 1992 I spent six weeks on a small island off the coast of Belize studying bottlenose dolphins. It was midsemester, and my fellow graduate students were busy with classes. My parents offered to join me because "they wanted to know what exactly it was that I did, and what my career as a scientist would involve." My parents had instilled an independent streak in me and my sisters; they taught me to have confidence without cockiness, to be strong-minded and inquisitive. It was okay to disagree with them in our regular dinner-table debates as long as we could defend our point of view. They supported me when I raised chickens in an urban community as part of my Future Farmers of America project, ran a traveling petting zoo with peers to teach kindergartners about farm animals, and kept a variety of odd, inedible items in containers in our refrigerator, just because. (My sisters were *not* thrilled with the last indulgence and lobbied our folks for an end to the container repository. As a compromise, I was relegated to the storage fridge in the basement.)

After my summer internship in Massachusetts, I began reading about dolphin and whale social and behavioral ecology—the research and periodicals section of the University of Connecticut library became my haunt. I began graduate school in 1990 under Bernd Würsig in Wildlife and Fisheries Sciences at Texas A&M University. After various modifications related to a change of study animal and location, I settled into a four-year stint that had me on a boat six months a year observing Atlantic spotted dolphins underwater in The Bahamas. This was not a vacation (but it was fun): every week, six to eight paying volunteers joined me to learn about and swim with these dolphins. All my fieldwork at each location where I have studied dolphins has been cosponsored by ecotourists. My early years of learning to work with inexperienced volunteers, learning in fact how to communicate my science with nonspecialists as I was becoming familiar with it, ingrained in me a passion not only for investigating dolphin communication but also for sharing what I learned with people.

The ultimate opportunity to share my research and the dynamic results of my colleagues in a public forum was offered about six months before I began postdoctoral research in Japan. I was in my parents' kitchen when I received a phone call from a film company based in California. While my mom listened to my side of the conversation, Teresa Ferrera spent thirty minutes asking about my research, results, tools, and details about dolphin social behavior. She told me that her company, MacGillivray Freeman Films, was planning to make a large-format film on dolphins for IMAX theaters. Would I visit their office to talk about dolphins to her boss and their production team, she asked. I'd been interviewed before for programs that never materialized. I said, "Sure, but I've just finished school, you'll need to cover the plane ticket and expenses." Four years later, _Dolphins_ opened in March 2000 in IMAX theaters nationwide. This forty-minute film is visually stunning and true to the science of dolphins. The work of seven scientists fills the frames, giving viewers the largest, most vivid picture of dolphin life. The National Science Foundation supported an educational lecture series for a fellow scientist in the film, Alejandro Acevedo-Gutierrez, and me to visit forty science centers and museums to complement the film. This experience was a once-in-a-lifetime chance to share my enthusiasm for science and dolphins with students of all ages.

In 2000, as *Dolphins* was premiering, I founded the Dolphin Communication Project (www.dolphincommunicationproject.org) as an umbrella for my research into dolphin communication and to provide an avenue for cultivating educational opportunities (such as ecotours, workshops, and seminars) that could inform the public. The mission of the Dolphin Communication Project is dual in its aim to foster opportunities for students in the sciences and to promote an awareness of marine mammal conservation. With my students (representing five universities) and through the project, I offer hands-on experiences during research expeditions and internships, fostering collaborative endeavors between scientific and educational programs.

I have been able to share my research, my passion for science and for dolphins with people throughout the United States, Japan, and Europe. I am humbled by how many people are openly and completely reverent about dolphins. When people find out that I study dolphins, they usually want to know all about my work, what I have learned about dolphin communication and intelligence, and what exactly I do. This is the link that we can foster with communication to promote conservation of dolphins and their ocean home. People will only protect what they love; they will only love what they know. It is our job to teach them, to inform them, and to pave the way for assistance.

Toni

I began to study dolphins a little more than twenty years ago. I am often asked how my career began, which takes me back to my childhood. That's when I began my quest to unravel the mysteries of the animal mind and heart. As a toddler, I was enthralled with animals, even those considered by others to be bothersome or boring. As a teenager, my first experience with dolphins came on a visit to an amusement park. My friends and I stopped at the dolphin tank and watched a performance, after which everyone turned to leave for the roller coaster. "Wait!" I cried out. This was my first chance to see "live" dolphins do what they do in their free time, not performing tricks requested by trainers during a show. My friends grew restless watching the dolphins slowly mill about, and moved on to the rides. To me, however, nothing in the park compared with watching these beautiful, lustrous animals. Enthralled, I watched the dolphins for the rest of the day and caught up with my friends later. Only years later did my enthusiasm for that experience wane when I came to know enough about dolphins to question the practice of keeping these keenly aware animals confined in a raucous amusement park tank.

As a teen, I caught the end of a television show featuring a biologist studying dolphins and seeking ways to communicate with them. Something clicked, and I knew that I wanted . . . no I *had* . . . to study dolphins for the rest of my life. I am usually methodical about making decisions, so this sudden career declaration was hard to explain, let alone justify, to myself. I knew I needed to begin a course of study. I found books written by the pioneering researcher John C. Lilly and learned that his foundation was located near where I lived in Los Angeles. I also began to read the works of other researchers, especially Kenneth Norris and Louis Herman. When I wasn't in school, I was volunteering for Lilly's Human-Dolphin Foundation and Marineland of the Pacific, where I worked primarily in rehabilitating stranded marine mammals. I also assisted in a boat- and land-based census of wild bottlenose dolphins along the southern California coast. Lilly's foundation invited me to participate in research on dolphin-human communication with Joe and Rosie, two captive bottlenose dolphins who had just been moved to the Florida Keys. After only a few days of working with them, I changed my plane ticket home to one with no specified return date.

My experiences with Joe and Rosie were fascinating. Also of interest were the other dolphins housed in the facility next to them. Every day, I watched dozens of people pay to get into the water and swim with those dolphins. Little did I know that I was witnessing the first commercialized "swim-with-the-dolphin" program. This also gave me a unique perspective on human impacts on dolphins and led me eventually to conduct the first studies of these swim programs. Toward the end of my stay in Florida, Joe and Rosie were being "untrained" in preparation for their reintroduction to the wild. The research project was ending.

At the conclusion of this life-changing summer, I received an incredible opportunity to live on a boat in The Bahamas for two weeks to interact with and study spotted and bottlenose dolphins in the wild. To see dolphins underwater, interacting not only with people and one another but also with the many natural features of their underwater environment, was an education beyond belief. I felt as though I had come home. Auspiciously, dolphin scientist Denise Herzing was on the boat. She was initiating research that would eventually become the longest underwater study of individual dolphins in the wild.

I began graduate school missing the dolphins and the crystal blue waters. Yet I was charged with renewed enthusiasm and had the good fortune to have respected researchers in the field as my academic advisers. So began my unique specialization in dolphin-human interaction and communication, as well as in behavioral indicators of dolphin stress in both captivity and the wild. I was encouraged, yet cautioned that academia would be critical, even cynical about my study of dolphin-human communication. Pop culture had of course romanticized this subject in earlier decades. But the persistence of this attitude felt stale and charged with anthropocentric assumptions and a lack of scientific objectivity. I wondered if underlying the resistance to this research was the idea that it was sacrilege to study humans in a way that other social mammals are studied and that dolphins, not being human, did not deserve such attention. This did not seem a very objective approach to the study of animal behavior since we are animals, too. With my professors' guidance, I applied ethological techniques to analyze dolphin-human interactions and found the results even more exciting than the glamorized, media-enhanced accounts of these encounters. The study of the dolphin-human bond certainly did not require a lapse of scientific precision. If anything, the interspecies sociality between dolphins and humans was more brightly illuminated by it.

My graduate research consisted of conducting the first studies on dolphin behavior in the context of swimming with humans in captivity and then in the wild. Subsequently, I received numerous solicitations from international government and regulatory agencies, movie production companies, and nonprofit animal welfare, conservation, and environmental groups to evaluate a variety of captive and wild marine mammal behavior in order to provide recommendations regarding their welfare, conservation, and management. I studied dolphins used in swim programs, petting and feeding programs, various species of captive and free-ranging groups of dolphins, as well as solitary sociable bottlenose dolphins, belugas, and orcas. My career has taken me to regions of the world that I never expected to visit, much less work in, and introduced me to a rich diversity of people ranging from animal protectionists, fishers and hunters to celebrities and prime ministers. The creation of a nonprofit organization, TerraMar Research (www.terramarresearch.org), gave me an independent yet structured organizational container for this profession.

When people ask me what is unique about my work, I reply that I feel privileged to study the complexities of dolphin communication and the effects of human interaction on so many different species under such varied conditions. I hope that my research helps to validate and encourage the scientific study of interspecies communication and the psychological and emotional lives of other animals. One aspect of my work stands high above all else: to conduct research that makes a positive contribution to the lives of our nonhuman kin. This is not only my scientific responsibility but a great honor. My goal is for people to appreciate dolphins for *who* they are as individuals, not just *what* they are to us.

Exactly how we met is hard for us to remember. Our first conversation was probably an excited discussion about dolphin communication, probably punctuated by a mutual dissection of the methodology used or the conclusions reached in some scientific paper we were both reading. In 1990, we were fledgling graduate students at Texas A&M. Under Jane Packard's expert direction, we participated in lively debates about ethology. Though we hailed from opposite coasts—Kathleen from the east, Toni from the west—we shared a mutual obsession for learning the mysteries of dolphin communication as well as a passion for studying dolphins underwater. Kathleen's interest in delving deeper into the study of dolphin-dolphin communication and Toni's curiosity about the scientific analysis of dolphin-human communication eventually evolved into our respective doctoral research projects. In graduate school, we occasionally assisted each other in the field. We have collaborated on many papers. The complementary nature of our differences and our similarities is reflected in our writing, which we hope contributes to the value and enjoyment of this book.

We are thrilled that dolphin communication is swiftly becoming a topic of choice among young scientists. There is much to learn. The cutting-edge research on dolphin communication of such early pioneers as Melba and David Caldwell, William Evans, Louis Herman, John Lilly, Kenneth Norris, Karen Pryor, William Schevill, William Tavolga, and William Watkins has inspired us and informed our collaboration with contemporary cetologists. Although we, too, have pioneered aspects of study into dolphin communication, we do not work in a vacuum. Science is an amazing journey; competing explanations, or alternative hypotheses, are often presented to describe one behavior. Discussion and debate infuse a field already flooded with questions. Readers will encounter examples of competing hypotheses in the following chapters.

To cover all points of view and provide an exhaustive review of dolphin communication would require several volumes of text and many years to write. In fact, we could not help but include some information from such related aspects of dolphins as their anatomy, evolution, conservation, and cognition. Still, we adhere to what we call the "Umi shopping analogy." Umi is Kathleen's "mighty sea beagle," and she loves rawhide treats and squeaky, stuffed animals. Kathleen delights in shopping for new toys for Umi but brings home only one toy at a time and only on special occasions. Umi is not aware of all toys from which Kathleen could select, but she is thrilled with each new arrival. In a similar way, we focus on the explanations and hypotheses we have come to know and trust based on our experiences and data. We do not present a complete catalogue of the varying explanations of each dolphin action, vocalization, or interaction. Curious readers are encouraged to pursue the additional readings we recommend and draw their own conclusions. Similarly, as your guides for this journey, we focus on landmarks we have visited and are most familiar with. In this way, you will become privy to our individual experiences with dolphins.

Referring specifically to the animals of our book, we should clarify what we mean by *dolphin* because the name alone can be a source of confusion. If you ask a fisherman about the animal that looks like the television star "Flipper" when he is out fishing, he will likely answer "porpoise." In some regions, fishermen use the term *porpoise* to refer to all dolphins and porpoises in an effort to avoid confusion between the dolphin and the dolphin fish, or mahi-mahi. The interchanging use of dolphin and porpoise occurs when these two words are used more generally in conversation and is not meant to ignore the existence of the Phocoenidae, which includes six species of porpoise—individuals different in several morphological and physiological characters from dolphins. We use the term *dolphin* to refer to members of the taxonomic family Delphinidae, which consists of thirty-three species of dolphins ranging from coastal to pelagic and tiny to large. For example, killer whales (orcas) and pilot whales are dolphins. Use of the nickname "whale" refers to their length, which is greater than 24 feet (about 8 m).

Although we focus on research from the study of dolphins, beluga whales (also toothed cetaceans and members of the family Monodontidae) make a significant appearance. This is because of Toni's extensive work with these animals and the extraordinary similarities that have been observed in their behavior and communication, particularly in terms of interspecies communication. We also include information on other cetaceans to provide comparative examples or because details are lacking on specific topics in dolphins. For this reason, readers will encounter sperm whales (the largest of the toothed cetaceans), baleen whales, and other marine mammals, such as seals and sea otters. Elephants, parrots, dogs, ravens, lions, human children and adults, and other terrestrial and avian animals also occasionally join us through these pages into the world of dolphins.

Chapter I A Dolphin's Life

They say the sea is cold, but the sea contains
the hottest blood of all.

D. H. LAWRENCE, "Whales Weep Not!"

The sea churns with foam. From the surface emerge shimmering bodies as far as the eye can see. They are dolphins, leaping above the water, visiting our world for a moment before disappearing from view. Again and again they surface—thousands of sleek bodies pushing toward a destination we do not know. . . .

This was my first view of dolphins in the wild. I was a student intern on a whale-watching boat off Cape Ann, Massachusetts. The sight of so many dolphins moving in unison was breathtaking. I almost forgot to record data, but I recovered and thus began documenting dolphin behavior for what will probably be a lifetime.—Kathleen

---

Common dolphins can travel in schools of thousands, even tens of thousands.[1] These ocean mammals, true to their name, are found throughout the world's oceans, and people often see them in vast numbers. Yet our view from above the water's surface does not allow us to see or hear them in the three-dimensional ocean world in which they live.

When we see dolphins, it is typically like this: from a boat or from land, dolphins appear for a second or two, and the rest is left to our imagination. This can be exasperating for researchers, especially because dolphins exhibit what may be the most complex, unique, and intriguing communication systems of all animals. But, when we go beneath the ocean surface, where dolphins live, we can find out how they live, how they communicate and interact with others, and maybe even more about who they are as individuals. As we delve into the realm of dolphins, we learn that they are not just members of a group but unique individuals, each expressing a kaleidoscope of physical and behavioral characteristics.

Imagine for a moment that an alien scientist is examining an aerial view of people on a busy street sidewalk. The people all appear to look and act pretty much the same from this vantage point. Only if the alien approached and followed individuals would each person's physical and behavioral discreteness emerge. And if this observer did not understand our language or could not hear or see in the same frequency range as us, an understanding of our world would be even farther out of reach. As aliens to a fully aquatic lifestyle, we dolphin researchers are in a similar situation. It is not always easy to admit how little we know about dolphin life; then again, that is what keeps our work so perpetually inviting.

Dolphins live in an environment that is foreign, even hostile, to humans, making their study and collection of data on their activity difficult at best. What keeps us coming back for more is not only the quest to understand the puzzle of dolphin behavior and communication but the excitement of glimpsing the world of another social being and the possibility of understanding another type of mind.

We have both attempted to gain a broader and more representative picture of how dolphins live in their underwater world. Our research represents two sides of the same coin. Kathleen investigates how dolphins interact with one another, and Toni focuses on how they interact with human swimmers, divers, and boaters. Our work requires us to spend extensive periods in the water with dolphins, observing their actions using underwater video cameras, still cameras, and writing slates. Although we focus on slightly different topics, our observations complement each other. This is true whether we have observed different species or the same dolphins in the same geographic area.

In The Bahamas, for example, we have both studied free-ranging Atlantic spotted dolphins, which travel in much smaller groups than their "common" cousins. From above the water, we see them swimming side by side as they surface to breathe. The underwater scene, however, tells a different story. These aquatic athletes sometimes swim frantically around one another, coming together to surface and breathe in unison, only again to split off from one another to resume their underwater speed ballet. Tracking the same individuals underwater often reveals a scattered, zigzag pattern, which requires us to follow the same pattern with our eyes as well as our swimming! Dolphins are excellent mimics and move in effortless synchrony. It is hard to keep pace with them. If a dolphin does not want to be near you anymore, she or he can vanish in an instant. Toni has observed dolphins swimming with humans in a manner similar to the way they swim with one another. When a dolphin is patient, the human and dolphin movements can be almost perfectly synchronous except a swimmer's snorkel is breaking through the water's surface alongside a dolphin's blowhole. An interesting observation in this regard is that dolphins often seem to "accommodate" some human swimmers by letting the swimmers set the pace, even when the swimmers are much slower and less agile, and need air much more often.[2] Similarly, Kathleen has noticed that dolphin mothers often accommodate their calves in a similar manner. The calves need to stay near the surface longer because they surface more often than mom to breathe. As calves mature,

Atlantic spotted dolphins are about the length of Kathleen or Toni when we wear our fins. Large, fluid movements are often mimicked by dolphins when they swim nearby.

they develop the physique to dive deeper and to expel and inhale more air at the surface with each breath. But until they can do this, mothers often stay with their calves near the surface.

By merging our views from above and below the ocean surface, we forge the most complete picture of dolphin life. Looking at how they interact with humans and other species helps us understand how they communicate. Dolphins are highly social animals. They often care lovingly for their young for years and assist peers in distress in a manner exemplary even by human standards. Of course, there are exceptions: Kathleen has observed wild Indo-Pacific bottlenose dolphin moms who would not win any parenting awards for nurturing, attentive behavior. These mothers even had a higher calf mortality rate compared with other females in the group.[3] The complexity of dolphin society is evident in how they fight with one another, as well as how they play, mate, feed, rest, and communicate.

Dolphins have long fascinated humans and are the focus of myths and fables in many cultures. Yet modern research has shown that many of these fantastical-sounding tales could in fact be true. And the complexity of dolphin society makes the facts even more fascinating than the myths. Some aspects of dolphin life are amazingly similar to that of other social animals—both aquatic and terrestrial, wild and domesticated. Observing them underwater allows us, as terrestrial social animals, to perceive their world through our senses. We eavesdrop on their lives and compare and contrast what we find to other social species, including humans. It is a moment where land and sea mammals meet . . . if only for a short time.

With their modified anatomy and streamlined form, dolphins are supremely adapted to life in the sea. Like us, they are mammals: they breathe air, are warm-blooded, suckle their young, and have hair before birth. But all of these things occur differently for dolphins than they do for land mammals. Dolphins are cetaceans, members of the order Cetacea. The cetaceans—whales, dolphins, and porpoises—are the most highly evolved, fully aquatic marine mammals, and make up 2 percent of the 4,600 living mammal species.[4] Dolphins belong to the suborder Odontoceti, or "toothed" cetaceans. Toothed whales are divided into ten families grouped into three "superfamilies": Delphinoidea, or oceanic dolphins, porpoises, and monodontids

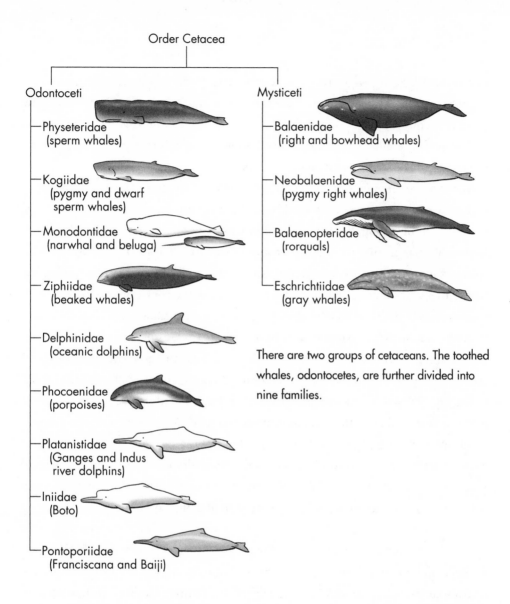

Order Cetacea

Odontoceti

- Physeteridae (sperm whales)
- Kogiidae (pygmy and dwarf sperm whales)
- Monodontidae (narwhal and beluga)
- Ziphiidae (beaked whales)
- Delphinidae (oceanic dolphins)
- Phocoenidae (porpoises)
- Platanistidae (Ganges and Indus river dolphins)
- Iniidae (Boto)
- Pontoporiidae (Franciscana and Baiji)

Mysticeti

- Balaenidae (right and bowhead whales)
- Neobalaenidae (pygmy right whales)
- Balaenopteridae (rorquals)
- Eschrichtiidae (gray whales)

There are two groups of cetaceans. The toothed whales, odontocetes, are further divided into nine families.

such as the beluga whale; Ziphoidea, or beaked whales; and Physeteroidea, or sperm, pygmy sperm, and dwarf sperm whales. The other suborder of cetaceans, Mysticeti, comprises eleven species of whales which have baleen plates instead of teeth.[5] Toothed whales also differ from baleen whales in having a single (instead of a double) blowhole, a highly specialized echolocation system, and a pronounced forehead, or melon.

Toothed whales make up the vast majority of cetaceans. Approximately seventy-one diverse species range from the relatively tiny vaquita, or Gulf of California harbor porpoise, which weighs in at about 120 pounds (about 50 kg) and is roughly 5 feet (about 1.5 m) long, to the well-known bottlenose dolphin, white beluga whale, and magnificent killer whale (the largest dolphin) on up to the largest toothed whale, the sperm whale, which can reach 55 feet (18 m) in length. The terms *porpoise, whale,* and *dolphin* are often used interchangeably, but size (specifically length) is the criterion anatomists have generally used to apply the common name *whale.* Porpoises, members of the family Phocoenidae, differ from dolphins in several characteristics. Typically smaller, they also lack a pronounced rostrum (beak) and have shorter, spade-shaped teeth as opposed to dolphins' more conical, pointy teeth. (Scientific names of species mentioned in the text are listed at the back of the book.)

Dolphins of the family Delphinidae are found in all oceans, most seas, and some bays around the globe.[6] Most species have a falcate (sickle-shaped) dorsal fin, cone-shaped homodont teeth, a pronounced beak or rostrum, and a gregarious social structure. They range in size from Hector's dolphin, about 4.5 feet (1.3 m) long, to the killer whale, about 30 feet (about 10 m) long. Dolphins are found in almost every aquatic habitat on the planet—from coastal to deep-water pelagic ocean zones and from fresh to marine environs. Of the thirty-three species of oceanic dolphins, the best-known and most studied is the bottlenose dolphin (think of "Flipper," of television fame). Bottlenose dolphins comprise the majority of captive dolphins in aquariums and are the species most often sighted along coastlines.[7]

As social mammals, dolphins and humans share many traits. We are both highly sociable and communicative, predatory, and intelligent, and we exhibit a variety of complex social relationships. Similarly, the cognitive abilities of dolphins are highly advanced.[8] For example, dolphins can recognize themselves in mirrors; only humans, some of the great apes, and elephants have been demonstrated to share this ability with dolphins.[9] In other words, although humans and dolphins lack a common ancestor, they have evolved similar cognitive abilities, possibly for comparable social or communicative reasons.

There is growing evidence that cetaceans have culture, similar to that observed in humans and other terrestrial (such as chimpanzees and elephants) and avian (such as

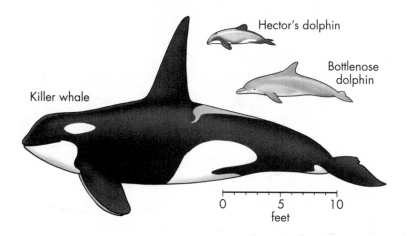

Killer whale

Hector's dolphin

Bottlenose dolphin

0   5   10
feet

The orca (killer whale) is the largest dolphin, and the Hector's dolphin is the smallest. The bottlenose dolphin is found in all oceans and seas, both near and offshore.

parrots, crows, and ravens) species. The complex vocalizations and behaviors studied in different killer whale populations are evidence of distinct orca cultures.[10] A form of cultural coevolution has also likely occurred between dolphins and humans in some regions of the world. For example, scientists have documented cooperative fishing on several continents between indigenous peoples and dolphins that appears to have developed over time across many dolphin and human generations.[11]

In an unexpected reversal of the typical path that humans and other terrestrial animals followed, dolphins returned from land to the sea about fifty-five million years ago. There are numerous hypotheses why cetaceans resumed an aquatic lifestyle. The most widely accepted are that the ancestors of modern whales, dolphins, and porpoises returned to water to take advantage of an untapped ecological niche, to use and more readily find a prey resource not being exploited by other species, and to have access to another food source without competition from other land dwellers.[12]

To get a sense of when and how dolphins obtained their unique and fascinating communication and related physiological systems, we must look far back in time. The cetacean branch of the tree of life starts out about fifty-five million years ago, when the ancestors of today's dolphins returned to the sea. From fifty-five to thirty-five million years ago, all ancient cetaceans are classified as Archaeocetes; at about thirty-five million years ago, the Mysticetes and Odontocetes—the two modern suborders of whales—first appeared.[13] If you traveled back in time fifty-five million years, you would not recognize those early cetaceans as being similar to their modern relatives. Rather, the proto-cetaceans were on a genetic trajectory that would take them, over fifty-five million years, to the species we recognize today. In fact, current theory suggests that one of the modern dolphin's ancient ancestors might have looked like a cross between a modern wolf, a cow, and maybe an alligator. Dolphins have had their current torpedolike, streamlined shape for about five million years. It's hard to imagine this, considering that modern human beings have been around for only about two hundred thousand years. Perhaps this extensive time for adaptation to the sea is one reason why dolphins have developed such a sophisticated system for sharing information—that is, communication.

Of the six families of Archaeocetes, two were the Pakicetidae and Ambulocetidae. The pakicetids, the oldest members of the lineage, were probably more semiaquatic than fully aquatic and had hind limbs that were close to fully intact. The ambulocetids, the second oldest family, also had hind limbs that were much more than "limb buds," but they were probably more aquatic than the pakicetids. The hind limbs of ambulocetids seem to be modified for an aquatic existence. As you move from fifty-five to thirty-five million years ago, and from pakicetids and ambulocetids through the other four families of Archaeocetes, there is increased reduction of the hind limbs and a more fully aquatic lifestyle.[14] Over time, about nine million years, these large feet became permanent flippers called a fluke, or tail, that was (and is) used to propel cetaceans through the water. The basilosaurids and

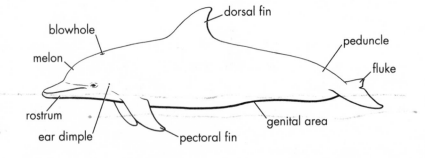

The anatomy of a dolphin readily shows its streamlined form.

blowhole

dorsal fin

melon

peduncle

fluke

rostrum

ear dimple

pectoral fin

genital area

dorudontids (not dinosaurs, though the names sound like it), now extinct, are the oldest ancestors of cetaceans to show solid evidence of fluke swimming.[15] Loss of hind limbs did not occur overnight but was a gradual progression as these animals adapted to an aquatic way of life; reduction of the hind limbs matched whales' and dolphins' need for functional locomotion in the water.

Anatomical streamlining for life in the sea included other modifications to the evolving dolphin body plan. Most mammals have two joints in the thumb (if present) and three in each of the fingers. Dolphins have several elongated digits per finger, even though the external view of their "fingers" more closely resembles a mitten. Imagine your hand in a mitten, but place your thumb inside with all the fingers, so that your mitten is more of a "mitt." A dolphin's flipper is homologous to the typical mammalian forelimb, for example to the human arm and hand, but it is covered by a webbing of blubber and skin that helps it function as a paddle to assist with steering through the water. Evolution has also imparted cetaceans with an elongated, or telescoping, skull, as well as a skull that meets the spine at 180 degrees, unlike the 90 degrees of most terrestrial mammals.[16] As the skull migrated from a right-angle connection, the cranial bones elongated, shifted, and migrated to different positions: for example, the nares (nasal openings) are now atop the skull, as opposed to in front of the skull, and act as a built-in snorkel. The dolphin's torpedolike shape reduces drag when swimming and indirectly helps reduce heat loss. In addition, modern dolphins show no distinction between the vertebrae of their lower spine, specifically the lumbar, sacral, and caudal vertebrae. They have significantly more vertebrae than most other mammals, which assist with trunk-generated, or axial, propulsion. This is a stark contrast to most mammals and is actually much more comparable to that of a snake.[17]

Molecular DNA studies confirm that the cetaceans' closest living relative is the hippopotamus.[18] Let's build some evolutionary perspective: think about hippos and their amphibious way of life. The hippo skull exhibits many traits in common with cetaceans: telescoping, upward positioned nostrils with flaps and a blubber layer under the chin for conducting sound.[19] Hippos can produce and receive sounds amphibiously, that is, both above and below the water's surface.[20] The hippo typically situates its ears, eyes, and nostrils above the water's surface, keeping its mouth and throat underwater. Researcher William Barklow distinguished nine categories of hippo sounds that were broadcast simultaneously in air and water.[21] Behavioral responses to other hippos and to playback experiments suggest that these social animals react to both the underwater and the in-air components of amphibious calls. They probably use sound to mark their territories and to mediate social confrontation. The ability of each hippo to know the location and concentration of other hippos in an area probably facilitates efficient grazing and use of pools.[22] A better understanding of how hippos use their sounds to coordinate social activity might provide insight into the mechanisms by which acoustic communication evolved in their modern-day oceangoing cousins.

Mesonychid condalarths are an extinct sister group of modern cetaceans. These wolflike carnivorous mammals inhabited coastal waters. Both physical appearance (morphology) and molecular data indicate ancestral ties among artiodactyls (modern even-toed ungulates), other even-toed ungulates like the mesonychids, or the modern-day cows or hippos, and dolphins.[23] The multichambered digestive system of dolphins and ungulates, such as the hippopotamus, elephants, deer, and cattle, are examples of overt anatomical similarities. Distinct links can also be found in their modes of communication, as we saw in the study of hippo sounds.

For many years, scientists did not know that dolphins vocalized in ultrasonic frequencies sometimes far above what humans can hear without the aid of electronic equipment. Early recording devices had a limited acoustic range. Only when recorders and hydrophones improved did researchers discover that dolphins were using ultrasonic frequencies. A similar discovery occurred with regard to baleen whales, many of which emit sounds far below what the human ear can perceive. So it was not surprising when biologist Katy Payne, a former whale researcher, discovered that elephants also vocalized outside the human hearing range.[24] In the 1980s, she and her colleagues discovered that elephants communicate in infrasonic frequencies that are so low that humans often cannot hear them, although infrasound sometimes can be felt by the body! This discovery both relaxed and intrigued elephant researchers, who had been mystified by the pachyderms' seemingly silent yet synchronous coordination of activities—even at distances of many miles. Elephants use sounds below 20 Hz to coordinate their movements and activities across great distances.

We find that blue whales do the same thing with infrasound over hundreds of miles, allowing communication across ocean basins. It is intriguing to consider the range of species using similar vocalization methods, each adapted to specific anatomical and environmental needs.

As it happens, what humans have learned about sonar (ultrasonic frequencies) and applied to medicine helped me (Toni) write some of this book. On a research trip in The Bahamas, I broke my foot. Undeterred, the next day I entered the water with my swollen foot. I was with about a dozen other snorkelers. Two spotted dolphins approached us and headed straight for me. I felt an intense "buzz" of echolocation, which vibrated through my injured foot then dissipated as it traveled up my leg. The dolphins then moved on to investigate the rest of our group, as if my foot was the only interesting thing about me. No one else indicated that they were echolocated on during that encounter. Even though the dolphin sonar did not miraculously heal my broken bones, I was healed through the wonders of modern technology with a device that helps heal bone using ultrasonic frequencies similar to dolphin sonar.

The dolphin brain has been the object of popular wonder, speculation, and, until recently, limited scientific research. Here we focus on the evolution of the delphinid brain. Given that there is a fifty-five-million-year divergence between cetaceans and their closest living relatives, one would expect brain structure to be quite different between the two. Research has confirmed that many extinct cetaceans had relatively small brains.[25] Over time, the cetacean brain increased in brain-to-body weight ratio and structural complexity, making them comparable to human brains in these two regards. Two significant evolutionary changes are often correlated with this advance. The first occurred near the origin of the Odontoceti from the Archaeoceti, about thirty-five million years ago. The second occurred approximately fifteen million years ago with the rise of the current Delphinoidae, the superfamily including Delphinidae and Monodontidae. Today, the absolute weight of the dolphin brain (when examined apart from its body) is a tad bit larger than that of the human brain.[26] In fact, when considering encephalization quotients (EQ), it is interesting to note that the dolphin EQ is second only to that of

humans in all the animal kingdom, including primates. Neuroanatomist Lori Marino and her colleagues have shown that the dolphin brain has a complex cortical cytoarchitecture, a point of comparison to the primate brain.[27] Dolphins require complex auditory neural processing to garner information from echolocation and other acoustic communication. This could explain their large brain size, but there is currently no evidence that dolphins have an abnormally large auditory area in the cortex dedicated to echolocation processing. Dolphins also have sleeping and waking patterns that are dramatically different from most terrestrial mammals and that likely require additional neural processing.

The evolution of an increased brain size in dolphins has been compared with the cortical complexity of living primates. There must have been good reason to evolve such a metabolically costly organ; otherwise, one would expect many more animal species to have large brains. Determining what the pressures were to drive this increase in cortical processing ability is the key to understanding why relatively few species have large brains. An increase in overall dolphin brain size and complexity seems to have conveyed a selective advantage for survival in the sea. Causes for the observed increase in dolphin brain size may be related to social ecology and communication.[28] Each dolphin taxonomic family, like those of many primates, is socially gregarious. Some species exhibit several convergent behavioral abilities such as mirror self-recognition, comprehension of artificial symbol-based communication systems and abstract concepts, learning, and the intergenerational transmission of behaviors (that is, culture).[29]

That both primates and dolphins have evolved large, complex brains is evident in our multifaceted social behavior and cognitive abilities. Why we have done this remains to be determined. Likewise, identifying how sophisticated, gregarious social behavior might be guided by complex brains remains to be examined. But the continued journey promises a rewarding outcome if we stay the course.

Dolphin biology permits dolphins to communicate—indeed to live and thrive—through a variety of sensory modes in a world foreign to the average human.[30] Their eyes, ears, respiratory system, circulatory system, skin, and blubber layer, to name a few, are all essential to their successful communication. Dolphin eyes are located on the sides of the head. Each eye has about a 180-degree field of view because dolphins can "pooch out" their eyes to the sides. This ability gives them limited binocular vision when they look below their chin. Around Mikura Island, Japan, on days with calm seas, I (Kathleen) have watched dolphins speed-swimming upside down just below the surface of the water. After a few observations I realized that the dolphins were tracking flying fish just above the water's surface. By traveling upside down and looking up, the dolphins could track the fish with their binocular vision. This foraging strategy seemed to ensure that the dolphins got dinner; when I was watching, the hunting dolphin never missed a flying fish as it reentered the water. Spherical lenses allow dolphins to focus both in air and underwater (human lenses are flattened).[31] Dolphins do not have tear glands, but they have other glands that secrete a viscous solution that seems to protect the eyes from the effects of saltwater submersion. This solution may also reduce frictional forces on the eyes at high swimming speeds.[32] One of the most unusual adaptations of the dolphin eye is its split, or double, pupil, which gives dolphins vision both above and below the water during daylight.[33] In bright light conditions, the two dark pupils might also help with depth perception.[34]

Can dolphins see in color? We're not sure. There is evidence both for and against a dolphin's ability to see colors.[35] Even though bottlenose dolphins have color receptor cells in their eyes, whether these receptors allow them to see color or simply enhance their ability to detect objects is unclear.[36] The color receptors, or cones, may actually function for increased visual acuity in low-light and bright light conditions.[37]

Species that live in relatively clear water need good vision, which is important for visual exchange of signals.[38] Visual communication occurs through use of an extensive variety of postures, positions, actions, and morphological features.[39] Various species have different color patterns. Atlantic spotted dolphins are born

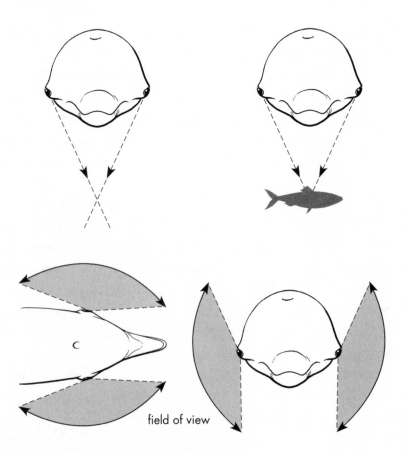

Dolphins can "pooch" out their eyes to gain more peripheral vision. Yet they have binocular vision (where both eyes see the same thing) only below their rostrum and throat. Each eye can see almost 180 degrees to its respective side.

field of view

without spots and gain pigmentation as they age, which allows for recognition of individuals as well as general age categories.[40] Killer whales show distinct black-and-white color patterns between their bellies and backs.[41] Dusky dolphins have a white belly that fades into gray sides and a black back, with a faded wide stripe or two for good measure.[42] Dolphins use their color patterns during direct and indirect communication with one another as well as when hunting or corralling prey. For example, foraging dusky dolphins will flash their white bellies at fish schools while cooperatively corralling them into a ball. With this action, a dusky dolphin is using several signals—posture, behavior, and morphological traits. While swimming in pairs, spinner dolphins will rotate and tilt their bodies toward and away from partners, using their pigment patterns to convey information.[43] The alternating presentation of the dark cape and white belly may allow synchrony among animals during dives and turns, especially when dolphins move as a group.

Dolphin skin is smooth with a rubbery feel. It is exceptionally sensitive, highly enervated, and well vascularized. Dolphins slough their skin regularly, and the outermost layer, or epidermis, is renewed every two hours, almost nine times faster than the human rate.[44] Dolphins are very sensitive to touch and are often observed touching or in tactile range of peers.[45] Dolphins may touch flukes, pectoral fins, teeth, and rostrums to a variety of places on the body of another dolphin. As in other mammals, dolphin touch can be an intimate communication between individuals or a form of play or aggression. Some forms of behavior are distinguishable only by the ensuing context. As a human analogy, think about when men exchange pats on the back: on the football field, a pat on the back might mean "congrats, nice catch"; in a dark alley, a pat on the back might be a prelude to a fistfight.

As mammals, dolphins must have hair, and they do, but only before birth. If you look closely, hair follicles are present on a dolphin's rostrum, but the hairs are usually gone. Lacking hair, dolphins would seem to have trouble staying warm in the water. The sea is a cold place for mammals, because water reduces heat from a warm-blooded body at least twenty times faster than air. Yet dolphins are able to maintain a healthy body temperature through a variety of adaptations: a decreased surface-area-to-volume ratio (accomplished through evolution by streamlining their shape); increased metabolic heat production (eating more); modified heat exchange

Resident killer whales are found in family groups called pods. All offspring stay with mom for life, with individuals recognized by their black-and-white patterns and family groups identified by vocal dialects.

systems (a circulatory system with radiator capabilities); and increased insulation (a blubber layer). Blubber not only insulates and improves dolphins' hydrodynamic shape, but provides a storage facility for energy during periods of fasting. The lipid content of the blubber layer directly affects thermoregulation; dolphins found in colder climates have a higher lipid content as compared to species with a tropical distribution.[46] Thus, for retaining heat, the actual thickness of the blubber may not be as important as the quality of the fat layer.

Dolphins are thus well suited to staying warm in the sea, but can they get too hot? And, if so, how do they counteract their ability to retain heat? The answer lies in a mechanism called a biological radiator. The circulatory system of all marine mammals has a striking feature—a *rete mirabilia* (Latin for "miraculous network").[47] Though many animals have retia, these systems are particularly advanced in deep-diving marine mammals. The retia function as countercurrent heat exchangers, or radiators. These radiators maintain a heat differential between oppositely directed blood flows, which increases the amount of heat transferred. Heat is retained in the core around vital organs in cold conditions and transferred through the blubber and skin in warmer climates. These radiators are present in the flippers, dorsal fin, flukes, and genital area. The retia keep the testes and the fetus cooler than the rest of the body, a key factor for species reproduction and survival.

Dolphins can hold their breath for up to fifteen minutes and are streamlined for efficient, swift movement through the water column (a conceptual column based on the horizontal layering of properties within the water).[48] When a dolphin dives, oxygen is shunted to smaller capillaries, the bloodstream, and the muscles. Oxygen is bound to myoglobin, a protein that binds four times more oxygen than does hemoglobin. Myoglobin is found in higher concentrations in the muscles related to locomotion, specifically in areas that produce greater force and consume more oxygen during swimming.[49] The increased amount of myoglobin present in marine mammal muscles makes possible longer and deeper dives by dolphins as compared with land mammals.

Studies on dolphins' swimming skills and deep diving capabilities have revealed that they do not actively flex their tail and back muscles when descending to or ascending from depth.[50] The dolphins use pressure changes and gravity differences between shallow and deep water; they kick a few times and simply glide down or up, depending on their direction of travel. This "kick and glide" method allows dolphins to conserve energy and still dive deeply for long periods, particularly for foraging, socializing, or investigating something in their environment. Many scuba divers and snorkelers follow the same kick and glide movements with their initial descent during a dive as well as during their return to the surface.

Speaking of kick and glide naturally reminds me (Kathleen) of the power available in a dolphin's peduncle, or tail stock. Mystic Aquarium, in Mystic, Connecticut, has a marine mammal rescue program through which dolphins that have come ashore are retrieved and, if possible, rehabilitated. One time when four Atlantic white-sided dolphins were stranded, I was helping to restrain a subadult (teenage) female being given fluids and food. My job was to hold down the spot where the flukes meet the peduncle—a dolphin's business end. I was kneeling on the flukes with one knee on either side of her peduncle with my hands holding tight to the peduncle just in front of my knees. Not your typical kneeling posture, but it served the purpose, or so I thought until she decided she was not happy with the treatment. As she was receiving fluids, she lifted her fluke, and me, about a foot off the ground! I remember balancing on the narrow peduncle of that dolphin while trying to get the attention of the other volunteers to add weight to her fluke. It was quite comical in hindsight, no pun intended. As never before, I understood the power of a dolphin tail, whether in or out of the water.

The next time you visit the ocean, take a deep breath and dive down for a few seconds. You can use the kick and glide method to save energy. You will hear that the sea is far from silent. The underwater arena is noisy, and humans are ill equipped to identify where an underwater sound is coming from. We have evolved to hear on land and to obtain directionality to sounds and their sources on land. Sound-source directionality is determined by an internal calculation in which your brain compares the arrival time of the sound at each ear. Because water is far denser than air, the perception coefficient is quite different between the two. Sounds travel about four and a half times faster and farther in water than in air. Therefore, to locate a vocalizing dolphin underwater, humans would need to do one of two things: increase the size of our ears by four and a half times or increase the distance between both ears by this same amount. Dolphins, however, are perfectly adapted to the underwater acoustic environment. They can hear and produce sounds from about 1 kHz (kilohertz) to more than 120 kHz, depending on the species.[51] In fact, dolphins have hearing and sound production capabilities that well exceed those of humans.[52] The human hearing range is roughly between 100 Hz and 20 kHz.

Mountains, deserts, grasslands, coastal regions, cities, suburbia, and the plains are just a few of the varied habitats that humans call home. Similarly, our oceans, seas, and even a few rivers offer an ecological multitude of options that dolphins might call home. Some species are littoral, spending their lives near a coastline. Others are pelagic, visiting the shallow coastline regions only occasionally and instead preferring the open ocean, far from the beach. Dolphins are almost always found in groups. Littoral species are found in groups ranging from five to ten individuals, whereas pelagic species are found in herds numbering in the hundreds or thousands.

In the past decade or so, cetologists have begun trying to correlate dolphin social life, behavior, distribution, and group size with their habitat. Bottlenose dolphins are found in all oceans and seas, along coasts and offshore, so they encounter a wide range of habitats and prey items, and other environmental conditions. Bottle-nose dolphin feeding strategies vary according to the type and distribution of prey, which in turn is related to habitat.[53] Near Hilton Head, South Carolina, bottle-nose dolphins engage in mud bank feeding.[54] In this method, several dolphins will coordinate to chase a school of mullet onto the shore; there is much splashing and chasing and lots of fish jumping into the air. The dolphins beach themselves on their right sides and grab fish in their teeth before shimmying back into the water. In Shark Bay, Australia, a few bottlenose dolphins also practice beach hunting and seem to learn the process from relatives.[55] In Golfo Dulce, Costa Rica, Alejandro Acevedo-Gutierrez observed bottlenose dolphins chase and capture fish (such as yellowtail jack) that were too large for them to swallow whole. As he watched, a dolphin would grab a fish by the tail and slam it against the water's surface, breaking the fish into smaller, easier to eat pieces.[56] Indo-Pacific bottlenose dolphins around Mikura Island feed on squid and small mackerel in the deep, productive water between Mikura and Miyake Islands. Closer to shore, these same dolphins snack on *takabe* and *tobiuo*. Takabe are a small schooling fish that the dolphins chase and capture regularly. Tobiuo are flying fish and require coordination and skill to catch: a dolphin will swim belly up just below the water's surface, paralleling the flying fish's flight pattern; when the fish tires or runs out of a sufficient air current and drops back into the sea, the dolphin is waiting with open jaws.

Although all dolphins are social animals, all dolphin societies are not the same. The smaller groups of littoral species merge and split more frequently when forag-ing, playing, socializing, traveling, and resting. Water temperature and depth, prey type and distribution, underwater visibility, and ocean currents are just a few of the variables that can differ for each species. The four species of freshwater dolphin in South America and India, for example, live in riverine systems that are often highly polluted, filled with particulate matter that reduces underwater visibility, and subject to heavy boat traffic. These factors, coupled with the dolphins' aloof, shy behavior, inhibited scientific study until the past ten to fifteen years. It turns out that Amazon river dolphins use a lek-mating strategy reminiscent of some ter-restrial hoofed mammals and birds, in contrast to the more promiscuous lifestyle exhibited by many oceanic dolphins.

Dolphins are usually found in groups ranging in size from three to four individuals to hundreds or thousands depending on the species and location.

Habitat not only affects a dolphin's dinner time, menu choice, and even mating strategy but also how each dolphin behaves as an individual, at least for some actions and signals. For the past ten years, I (Kathleen) have been investigating how dolphins use their pectoral fins to share tactile signals. Why? I'm interested in deciphering whether static contact and rubbing with a pectoral fin have different meanings.[57] I am also curious whether contact with different body parts has different meanings and whether initiator and receiver roles affect the meaning of flipper contact. While conducting research, my students and I discovered that Indo-Pacific bottlenose dolphins around Mikura Island, a dormant volcano with a rocky boulder shoreline, use pectoral fin contact in subtly different ways than do The Bahamas' Atlantic spotted dolphins, which live over a white, sandy ocean floor. We often see spotted dolphins rubbing their bodies in the sand—maybe to scratch a nagging itch or slough off parasites. But the bottlenose dolphins at Mikura don't do this. In the interest of science, as well as to gain a dolphin's perspective, I rubbed my hands and arms on both bottom surfaces. As you would expect, the sand was smooth and soft, almost feathery, whereas the rock surfaces were so jagged and rough as to be almost untouchable without damaging my skin. I can fully understand why the Mikura dolphins rub their bodies on the rocks only when a thick, leathery mat of *tosaka* seaweed covers the surface. Thus, it also appears that the environment, in this case the sandy versus rocky bottom, might affect the expression or use of a particular behavior. Think about this: if you have an itch in the middle of your back and you are alone, then you might rub your back on a doorjamb or similarly smooth object. But, if someone is nearby, you can ask that person to scratch your back.

As we've mentioned, many dolphin species are gregarious, with both consistent and fluid social relationships.[58] Bottlenose dolphin groups usually average ten to twenty-five individuals, but the gender, relative age, and identities of a dolphin's companions often change throughout an individual's life.[59] Many small dolphin species' societies have been compared with those of chimpanzees in having a fission-fusion dynamic. That is, dolphins spend a lot of time in small groups traveling, foraging, and playing and come together to form larger groups for socializing, coordinated foraging, and other activities. When finished, dolphins break back into smaller groups with the same or different membership as before. There are exceptions to this societal format in some dolphins. Larger dolphins lead somewhat different social lives. Bottlenose dolphins in Sarasota Bay, Florida, and in Monkey Mia, Shark Bay, Australia, have been studied for several decades and exhibit the classic mother-calf relationship, in which the young remain extremely close to their moms for several years. Bottlenose dolphins from Florida and Australia are known for the strong, durable friendships that male dolphins form, often lasting a lifetime.[60] These male pairs form coalitions during the mating season and either coordinate or compete for access to breeding females.

Killer whales have a matriarchal society—young orcas stay with mom for life. Long-term studies have revealed three distinct populations of killer whale in the northeastern Pacific, the waters of British Columbia, Washington, and southeastern

Alaska. Two different populations, or ecotypes, inhabit the same coastal waters, the "transients" and "residents," yet they remain socially and genetically separate, differ markedly in seasonal distribution, social structure, and behavior, and even exhibit such different physical traits as the shape of the dorsal fin.[61] The third orca ecotype typically remains offshore, and little is known about this population's diet or behavior.

The behavior of free-ranging dolphins reflects a dynamic interplay of aggression, social and sexual interactions, alimentary and exploratory behavior, play, flight and predator avoidance, and assisted locomotion.[62] Aggressive behavior among dolphins is not uncommon and may occur along with sexual behavior.[63] In addition to reproduction, sexual behavior is involved in social bonding and dominance.[64] Care by the mother or other assisting subadult or adult is frequently documented in most dolphin groups.[65]

Dolphins play throughout their lives, although juveniles play significantly more than adults. Dolphin play is defined as any behavior that is not directed toward the satisfaction of hunger, travel, or other biological requirements.[66] Although scientists once held that free-ranging dolphins do not play as much as their captive counterparts, this conclusion was based on limited underwater observations of free-ranging dolphins.[67] Atlantic spotted dolphins in The Bahamas and Indo-Pacific bottlenose dolphins around Mikura Island engage in daily play.[68] The motivation to play in dolphins is in fact so strong that play with objects has been used as a sole form of positive reinforcement for training captive dolphins, as well as a form of environmental enrichment for dolphins when not interacting with their trainers.[69]

Play behavior serves many functions in social groups: asserting, determining, and establishing relationships and perhaps self-pleasure and development of motor skills.[70] Play activities vary greatly in both form, function, and frequency relative to season, gender, individual differences, species, and social and environmental contexts.[71] In dolphins, mimicry, manipulation of objects, chasing, bow-riding, and different types of jumping, turning, and tactile interactions have all been identified as play.[72] An increase in motor activity that does not provide any other apparent benefit has also been identified as a primary characteristic of play behavior in dolphins.[73]

Rubbing flippers (pectoral fins) can be a greeting between dolphins that have reunited after a short time apart.

The study of marine mammals is a relatively new field. Studies of dolphin anatomy, physiology, distribution, population dynamics, ecology, and behavior have been ongoing for almost a century. Yet systematic quantitative data on delphinid behavior and communication have accumulated only within the past forty years or so.[74] Data are gathered from a variety of viewpoints: surface observations from boats (including whaling vessels), land-based stations (such as cliff tops), aircraft (planes, hot-air balloons, blimps), and captivity and, since the early 1980s, underwater observations. Detailed observations are conducted on about a dozen of the more than twenty genera of dolphins (oceanic and freshwater) and porpoises that have been identified in the wild. Further, almost all of these studies have been conducted on coastal or near-coastal populations.[75] Coastal bottlenose dolphins, harbor porpoise, and short-finned pilot whales are regularly studied in many geographical areas.[76] Atlantic spotted dolphins in The Bahamas, Hawaiian spinner dolphins in Hawaii, dusky dolphins near Argentina and New Zealand, humpback dolphins off South Africa, and killer whales in several locations are currently being investigated extensively.[77]

With the exception of the studies of Hawaiian spinner dolphins, Atlantic spotted dolphins, and Indo-Pacific bottlenose dolphins at Mikura Island, most intensive studies have been conducted from above the water's surface. Although many studies have been conducted on dolphins in captivity, results have not yet been extrapolated to free-ranging populations.[78] Similarly, studies of dolphins in the wild can provide insight into observations of dolphins in captivity. In either setting, observing dolphin behavior is invigorating and thrilling for each of us. Imagine starting your day with the sights and sounds of dolphins instead of a cup of coffee.

In remembering an ideal dolphin day . . . I (Toni) recall being awakened by the sound of a research assistant whispering loudly, "Dolphins!" With the sun barely up, I throw on my bathing suit instead of my bathrobe. As second nature, I grab the underwater housing with camera prepared for action the night before. After placing a mask over my barely open eyes, I slip quietly into the ocean. This is my favorite way to study dolphins as well as to start my day. When waking up to dolphins, my mind becomes quickly alert from the adrenaline that, even after decades, provides a rush of amazement whenever I see these marvelous beings.

One morning, before entering the water, I noted lots of spinner dolphin mothers and calves close to shore. Although I wanted to observe their underwater behavior, I also didn't want to interrupt their rest or other life-nurturing interactions that might be occurring. So my fellow swimmers and I intentionally swam away from the group, hoping to record vocalizations and watch them from a distance. I heard nothing, at least nothing within my hearing range. But I saw something approaching slowly underwater: at least half a dozen female spinner adults, each swimming with a fairly small calf. I quietly got the attention of my human pod and we dove slowly down together. The dolphins were moving so slowly that we could barely detect body movements from the adults, though the little ones propelled their flukes more noticeably. In our minds, the mothers seemed to be bringing their young to introduce them to us. They glided below us and with precise synchrony angled themselves to look at us with their left eyes. As I videotaped, I intentionally angled my body toward them in reflection of their posture, hoping they would sense that I was mirroring their actions. We swimmers found ourselves moving in synchrony almost effortlessly with one another and with the dolphins for several more dives, all slowly gliding together punctuated by the sound of almost simultaneous snorkel and blowhole breaths and the occasional click of a camera. To me, this was a spontaneous dance between humans and dolphins.

I (Kathleen) often feel the same when I watch dolphins interact with graceful ease through the water column. While making the movie *Dolphins* for IMAX theaters with MacGillivray Freeman Films, I used an underwater scooter to keep pace with a group of spotted dolphins. The feeling was exhilarating. For the first time, I could keep up with dolphins that arced through the water and spun on a dime. I led them in turns twice before I again was the one lagging behind, but even though I was not the leader of the pod anymore, the dolphins hung out with me for about two hours. Invigorated, excited, thrilled, awed, mischievous, and, ultimately, exhausted are only a few of the feelings that coursed through me that afternoon. I was invited to the dolphins' dance and delighted by kicking up my heels for fun.

Our accounts represent more than encounters between humans and dolphins. Within each of us there is also an internal dance between two selves: one self is the dedicated, objective researcher who documents every nuance of dolphin posture, behavior, interaction, vocalization, and more with exacting precision; the other is the little kid who is simply enamored by dolphins and wants to cavort with them until the sun sets. As scientists, we have both a yearning and a responsibility to share something more than our personal experience of these incredible animals of the sea with the terrestrial world. Fortunately, over the years, we have learned to unite these different selves. Our desire to "join" the pod and play has been refined so that, rather than hindering our observations and subsequent analyses, it actually enhances and supports it.

Chapter 2 The Expressive Dolphin

It is of interest to note that while some dolphins are reported to have learned English—up to fifty words used in correct context—no human being has been reported to have learned dolphinese.

ATTRIBUTED TO CARL SAGAN

Esteban had Alita in his sights. She was the newest female in the group and the only one who was very pregnant. Try as he might to wiggle his way past the other females of the group, they simply would not let him pass. Esteban's frustration was mounting, as evidenced by his whistles, clicks, and jerky movements around Gracie, Rita, Mrs. Beasely, GeeGee, and Cedena—Alita's elite guard against Esteban's amorous intent. Suddenly, Esteban did a sharp U-turn, stopped vertical in the water column not more than 6 feet (2 m) from me, and began clapping his jaws together in my direction! It is not every day that a large male bottlenose dolphin jaw claps at you. Jaw claps from adult males are often considered a sign of aggression, a distinct signal to "back off" or "get out of the way." My reaction could not have been more opposite: I kept filming and did not budge. Esteban jaw clapped for about thirty seconds and then zoomed out of sight. Hindsight had me rethinking my response, and I have revisited the action in my mind. Esteban was so frustrated, I think, that he needed to express his irritation at the nearest being, me. That is, he seemed to be "venting" about his situation before leaving to pursue some other behavior.[1] In Esteban's situation, his jerky movements and vocal signals were likely intended to send a message to the adult females barricading him from Alita. He wanted access to Alita. But could the same be said for his jaw claps at me? Was Esteban trying to send a message to the first being he encountered after being rebuffed? Or was he just "letting off steam"?—Kathleen

Communication is the cement that binds all animal societies. Every day we share information with family members, friends, and business associates, not to mention numerous strangers in stores, gas stations, and other places. I (Kathleen) believe that Esteban was trying to communicate with Alita and the other female dolphins, though I cannot say the same for his actions toward me. That is, Esteban's jaw claps seemed more like the phrases my husband utters in my direction when his attempts to fix a "sensor error" on our car go awry. This is more "self-communication" or expression than an attempt to share information with another being.

Conversely, I (Toni) have interpreted this same dolphin jaw clap behavior some-what differently, as a threat or warning, albeit in different situations than that which Kathleen describes. I have observed instances of dolphins jaw clapping at another dolphin or person, though I have not yet been the recipient of this behavior. In a few of my observations, the jaw-clapping dolphins also directed forceful, intentional physical impact (such as biting or ramming) toward a recipient, that is, the other dol-phin or person. When these aggressive behaviors followed jaw claps, the recipient did not retreat or back down (which I think makes Kathleen a lucky person). I have also seen a young wild beluga clap her jaws toward a human swimmer who didn't know better, thought the behavior was "cute," and did not retreat. If the behavior was a threat, the beluga did not follow through, but she did leave the area shortly thereafter. So perhaps when signals are meant as threats, the threats are not always followed through . . . or, maybe they are bluff and bluster. And as Kathleen suggests, sometimes dolphins that exhibit this behavior are really exhibiting frustration or stress rather than conveying an intentional signal. The question remains, how do we as observers, and other dolphins, know the difference?

What would life be like if we could not communicate our wants and feelings to others? It would certainly be immeasurably more complicated and difficult, if not impossible. Social behavior is organized into patterns of coordinated behaviors and activities. To understand communication in a particular animal or species, we need to learn what, how, when, and why certain signals are used. In other words, we must discover and analyze as many basics of animal communication as we can get our hands, eyes, and ears on.

Some basic building blocks will help clarify the nuances of communication and the study of how other social animals use information to survive the murky waters of society.[2] Three pieces of information are critical in the study and comprehension of communication: the signal, the signal sender, and the signal receiver. The signal is the vehicle (the "what" and "how") by which the sender and the receiver exchange information. The sender and receiver must exchange signals accurately to avoid missing the message, which would cause miscommunication. Signals can be subtle, such as an animal's display of a unique color pattern or a slight change in body position. Or signals can be more dramatic. Think of a lion's loud roar or a dolphin slapping its tail to warn an intruder. Signals are often accentuated by other behaviors, and sometimes the sum may be more than the total of its parts. Play behavior in dogs and other animals includes bites and growls combined with play bows—without the play bow, the associated actions could perhaps indicate the opposite of play. When we see a growling dog with bared teeth, it might also have its tail straight and not wagging, and the hair along its back might be standing on end. Taken together, the meaning of these signals would be hard to misinterpret, even across species.

A signal can be modified based on location, distance between participants, past experience, age, sex, reproductive status, environmental factors, presence of peers, quality and history of relationship with others, and other factors, including the "mood" of the senders and receivers. Therefore, the meaning that a signal conveys relies on contextual cues that are crucial to imparting the correct message. A signal given in one context could result in a different meaning in another: a "jaw pop" from a dolphin might be used to threaten an intruder in one situation or reprimand a youngster in another circumstance.[3] A dolphin's leap out of the water may signal to other dolphins that fish are nearby whereas a slightly different breach may signify play. A signal conveys both a message and a meaning—the nuts and bolts of a communication exchange, so to speak. The message is the information provided by the sender, whereas the meaning is the significance attached to that signal by the receiver and varies by context.

Many species throughout the animal kingdom are capable of sophisticated communication. Through what is referred to as a "waggle dance," bees signal the presence, abundance, and location of food to other members of the hive.[4] Researchers Joyce Poole and Cynthia Moss have noted combinations of elephant matriarch vocal rumbles associated with actions that appear to communicate the direction of travel for the group. Poole suggests that these vocalizations may constitute a word as defined by primatologist Sue Savage-Rumbaugh.[5]

When we study communication, the signal is the most discernible part of the exchange, but how the receiver perceives and responds to a signal is just as important. The message and meaning of a signal are not always the same; different receivers may acknowledge very different information from the same signals, and contextual cues play an important role. To give an example in human communication, consider walking through an airport terminal. You see people waving to or hugging other people. Are they saying "good-bye" or "hello, welcome home"? The mechanism for miscommunication is similar (same bolts) yet different (nuts). For example, a man who is wearing a new suit hears a good friend tell him that he looks "unusually handsome" and takes it as a compliment; in contrast, if a competitor at work says the same thing to him, he may interpret the remark as an insult—even if it was intended as a compliment. The social context of the latter session could lead to misinterpretation. In evolutionary terms, these are everyday signals that humans use to strengthen bonds between friends and family members. Most signals have evolved to address specific problems or meet certain needs in social behavior, such as finding a mate and coordinating efforts in searching and capturing prey. Senders choose from multiple sensory abilities to encode a specific signal, while receivers use their own diverse sensory set to decipher the meaning of each signal.

Communication can occur via a single sensory mode or by a combination of multiple sensory modes. These can include visual, acoustic, gustatory, tactile, and olfactory signals. For humans, gestures or postures, such as standing with one's hands on hips or shaking a pointed finger at someone, can modify a verbal message or even convey one on their own. A handshake, hug, or pat on the back following a conversation often conveys more about the relationship between the individuals and their intentions than the conversation itself. Touch occurs with any body part, but some areas have special significance.

How might dolphins use their pectoral fins to share information? When one dolphin places its flipper on another's side between the dorsal fin and the tail, it may be requesting something. The actual "favor" being asked depends on the context, but the touching dolphin initiates the request with its flipper. To appease an excited youngster, adult dolphins will often rub the young animal's melon (forehead) with a flipper. Young spotted and bottlenose dolphins often swim excitedly around peers and humans, only to be curbed by an adult with the placement of a flipper on the youngster's head. In 1993, I (Kathleen) watched an adult female spotted dolphin that had been "babysitting" a group of youngsters keep her pectoral fin on her calf's melon for several seconds after a particularly swift game of keep-away with several other young dolphins. Flipper-to-flipper rubs appear to function as greetings like handshakes or hugs among people.[6] The situation often dictates the specific function or meaning of the flipper contact, be it a touch or a rub between individuals. Pectoral fin contact may also signal affection to another dolphin.

Individuals might use their flippers to touch or emphatically rub a body part of another dolphin. These flipper contacts are often modified by other signals, such as posture or vocalizations, to make the point even clearer. Mutual flipper rubs might be a signal to renew a friendship whereas use of a single pectoral fin by one dolphin patting a second may signify appeasement.[7] The latter behavior is especially evident to human observers when the sender is an adult and the receiver a younger dolphin. Compare flipper rubs with handshakes or pats on the back in humans: the meaning varies, even if slightly, depending on the context as well as the age and sex of the participants. So, messages can be sent by a variety of methods, and for many species, actions often speak louder than words.

One thing that has become abundantly clear is that dolphins use their flippers to exchange many types of messages with peers. But could use of the flipper to share information be extended to a human being? More specifically, could pectoral fin rubs be used by a dolphin greeting a person? From 1991 until 1995, I (Kathleen) spent each summer studying the communication patterns of Atlantic spotted dolphins in The Bahamas. I returned briefly in 1997 and then, after a two-year absence, resumed research there in 2000. Although it would be hard to test or prove scientifically, I secretly wondered if the dolphins would remember me. If so, how would I know? Many dolphins that I had identified previously had aged and gained more spots; some had offspring of their own. It felt as if my family was growing up! What surprised me most, however, was how closely the older ones—and their young born during the two years that I was away—approached me. A couple of dolphins even rubbed their flippers against my body! Had they remembered me? Had the young dolphins been told about the "two-legged being (me) with the green tube" (my camera system)? We have no way to prove this or even to ask dolphins what they mean when they rub a human. But considering that dolphins use flipper rubs and other rubbing contact as greetings in certain contexts, and considering that these dolphins rubbed me when I saw them for the first time that season, a meaning of "hello" or "nice to see you" might not be far-fetched. Of course, the rubbing could also simply have been an artifact of the increased tolerance and habituation to human swimmers by newer generations of Atlantic spotted dolphins. Something to ponder on the boat ride home . . . or when watching the videos over and over at the lab.

Research on dolphin acoustics suggests that these animals use tonal whistles of mid- to high frequencies (or pitches) primarily when socializing.[8] These whistles appear unique to dolphins, and they vary across species as well as within species across geographic locations, social groups, populations, and even individuals of different gender and age.[9] Pulsed broadband tones in the form of echolocation clicks are associated more commonly with nonsocial activities, such as foraging or short-range investigation.[10] Burst-pulse sounds—short broadband vocalizations—vary in intensity and encompass squawks, squeaks, moans, barks, groans, and yelps, to name a few.[11] Although they are used in social contexts, these sounds have not been studied as extensively as other dolphin vocalizations.

How have we been able to learn about the sounds that dolphins produce? Have we been able to decipher why dolphins use certain sounds for particular reasons?

Dolphins have been held in captivity since the 1800s, but not until the early 1950s were scientists able to document and record sound underwater: the first hydrophones were used by David and Melba Caldwell at Marine Studios in Saint Augustine, Florida, to document dolphin whistles and clicks. We have learned much since that discovery. Although dolphins have vocal folds (homologous to our vocal cords), they are modified and probably not used in sound production.[12] In the early 1990s, Ted Cranford, Mats Amundin, and Kenneth Norris used MRIs and CAT scans to study the dolphin skull and began to illustrate how dolphins might create their repertoire of sounds.[13] They examined forty individuals from nineteen species of dolphins, porpoises, and other toothed whales, generating the currently accepted hypothesis that toothed whales have a "biosonar signal generator" in the nasal complex located approximately behind and below the blowhole in the head. This biosonar signal generator (or MLDB complex) consists of two structures: the dorsal bursae and the phonic lips (nicknamed "monkey lips" because they apparently resemble the lips of a chimpanzee). The passage of air over these structures creates the pulses that produce echolocation, or dolphin biosonar, which has a signal at about 120 kHz—a value significantly greater than the sound frequency a person could either make or hear. The fatty acoustic lens tissue in the melon gives dolphins a way to direct these pulsed sounds. Scientists once thought that dolphins used echolocation only to detect prey and in short-range navigation and that whistles, or frequency-modulated pure tones, were the primary vocal signal used in dolphin communication. But not all dolphins whistle, which suggests that echolocation, as well as other pulsed sounds, might also be used in communication.[14]

As most people know, dolphins breathe through a single blowhole. Less well known is that they have two openings in their skull, just as terrestrial mammals do. Through evolution the left nostril has become the blowhole and the right has been modified into four sets of air sacs that reside above the skull, behind the melon, and below the blowhole. Dolphins are believed to make their whistles by passing air back and forth between these sacs, which is not to be confused with the MLDB complex, where sonar is produced. If this is hard to imagine, think of a balloon that you would inflate using your lungs. Blow up the balloon but don't tie off the end; instead, pinch the end flat and let the air escape. The resulting sound is a good imitation of a dolphin whistle!

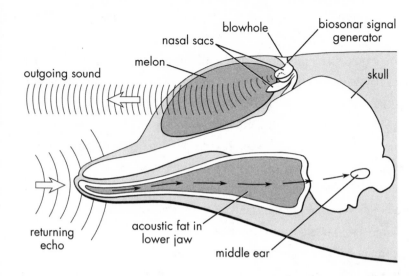

The dolphin melon acts like a lens and helps direct the pulsed sounds that are echolocation. Dolphins receive sounds (that is, they hear) through their lower jaw, which is hollow and filled with the same fatty lipid that makes up the melon. Whistles are generated by air being passed back and forth between four sets of air sacs behind and below the blowhole.

music?
food?

If an animal is going to use sound to communicate, then it should be able to receive and understand the acoustic signals used. Hearing, in its most simple definition, is the detection of sound. Technically, acoustic waves are the propagation of a mechanical disturbance through a specific medium; sounds are waves, just like light, but their structure is different. Think about a thunder and lightning storm. We see the lightning before we hear the thunder (unless the storm is directly overhead), but we also feel the rumbling of the thunder as it approaches. This is because we can feel sound waves over parts of our bodies, but we use our ears to capture the sounds, which are then transported to our brain for deciphering. Even though dolphins lack external ear pinnae, they have an internal ear bone structure at least comparable to most terrestrial mammals (including humans).[15] The middle ear is essentially isolated and insulated from the rest of the skull by the ecto-tympanic membrane and is connected to the skull by one small piece of cartilage.[16] This isolation reduces interference from the echolocation waves produced in the cranium. The tympanic membrane (eardrum) is attached to the ecto-tympanic membrane by a narrow membrane on one end and the hammer (malleus) on the other. It has the appearance of a folded umbrella, which is very different from that of land mammals.[17] The ossicles—what humans generally call the hammer, anvil, and stirrups of our middle ear—are also somewhat different in shape relative to those of terrestrial mammals, but the same general configuration exists.[18] The muscles of the middle ear are greatly reduced, having been replaced by a thick collection of veins and arteries. This plexus of blood vessels is probably an adaptation for deep diving. The plexus might be inflated to reduce airspace in the middle ear cavity, thus increasing pressure to match ambient pressure when at depth.[19]

If the ear is more adapted as a diving mechanism and we know that acoustic signals are one of the most efficient methods of sharing information in the oceans, then what might be the most likely way for a dolphin to receive sound? Here's a hint: although their smile is probably one of the most recognizable traits of all dolphins, it is also functional. Dolphins receive sound with their lower jaw.[20] Their lower jaw is actually hollow and filled with a lipid nicknamed acoustic fat. This same acoustic fat that fills the dolphin lower jaw also resides as a sort of lens in the melon, the dolphin's forehead. The lipid lens in the melon helps direct sounds passing from the dolphin to the environment. The lower jaw acts sort of like human outer ears to help the dolphin receive sounds and get information about the direction from which a sound originates. The overall blubber layer, which is great for warmth, streamlining, and energy stores, also insulates dolphins from being bombarded by sounds from every direction. Unlike dolphins, when we humans are in the water, we tend to "feel" the sounds more than actually hear where they come from. One reason for this is because sounds travel about four and a half times faster underwater than in air, and we are built to identify the direction of a sound source only through air. Over time, dolphins evolved to a social life in the sea; they have developed adaptations that allow them to take advantage of their environment and optimize its properties to facilitate information sharing—communication—among group members.

Why would sharing information be important to social life? As we mentioned earlier, a coordinated social life relies heavily on information sharing among individuals of a group. The ability to share information within a group requires complex social coordination that is learned at a young age. Think about our own species and how much play can resemble fighting from a distance. The behaviors of animals engaged in play or fighting are remarkably similar. Play is often characterized by mock fighting behaviors like biting, hitting, chasing, and growling or other loud, harsh sounds, but subtle cues remind those involved (as well as those observing) that the activities are not truly aggressive. Play in young animals benefits the formation of long-term social attachments; through play, the young are learning the meaning and proper use of the signals within their social structure.[21] Think of puppies or dogs, kittens or cats, or your younger brothers or sisters as they play and learn what is proper etiquette within the rules of the game. I (Kathleen) have often marveled how young spotted and bottlenose dolphins always seem energized by a good game of chase or underwater keep-away. I've watched a group of juvenile and subadult spotted dolphins play keep-away with a sea cucumber. They seemed like children on a playground zipping to and fro with their toy. Learning a set of communication skills, or acquiring knowledge of the signals appropriate for sharing information with peers about a specific topic, will benefit each individual as he or she grows and becomes a member of the adult community. The sea cucumber likely served as a bonding tool, or toy, for the dolphins during play that helped them establish their budding friendships.

Coordinated feeding behavior is one of the most important learned skills among dolphins that requires an efficient exchange of information. What dolphins eat as their primary prey often depends on where they live: dolphins generally eat squid, octopus, shrimp, some invertebrates, and fish. The bulk of a dolphin's diet is fish: dolphins hunt both solitary and schooling fish from within the water column, near or far from shore, from above the water surface, and deep in the ocean. Dolphins

forage for prey through a variety of strategies that require coordination and communication among individuals, including hunting cooperatively or in groups, crater
feeding (named for the crater depression in the sand after a dolphin pulls out a fish
that was buried there), chasing fish onto mud banks and beaches, stunning fish in
the shallows by whacking or kerpluncking them with their flukes, and corralling
fish with bubbles.[22] Beach hunting, also known as strand feeding, is one strategy
employed by killer whales and bottlenose dolphins in different geographic regions.
Individual members of some killer whale populations in the North Pacific, South
Atlantic, and Indian Oceans will beach themselves to hunt amphibious mammal prey
such as sea lions. In this method, the hunters patrol the shoreline from a moder-
ate distance while remaining silent. They strike with speed and agility when they
identify their unsuspecting prey. Adults teach this strategy to young killer whales
both by example and with direction: to this end, an adult may actually toss a caught
sea lion to a juvenile when class is in session. Transmission of this foraging tactic
is cultural and has been documented among generations of these leviathans.[23] The
young orcas must practice their beach approach as well as their shimmy to get back

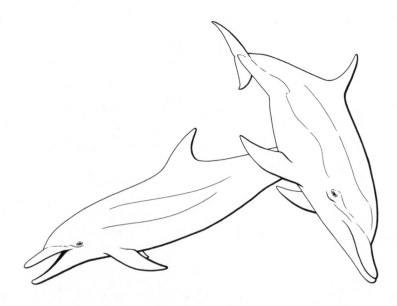

When dolphins fight,
there is no affectionate
contact or rubbing
action. They approach
each other at right
angles or head to
head.

When dolphins play, they will often "wrestle" over each other and approach each other at oblique or odd angles, even from behind. They will also rub each other often when playing.

in the water and avoid becoming truly stranded. As with many human skills acquired during youth, practice makes perfect.

Another form of beach or strand feeding is practiced by bottlenose dolphins along the East Coast and in Australia. Along the muddy shores of Georgia and South Carolina, bottlenose dolphins will chase mullet onto the beach by creating a pressure wave that the fish ride and from which they leap.[24] In Perth, Western Australia, however, only four adult female bottlenose dolphins and their calves from among the larger population feed in the shallows along the Cape Peron beaches.[25] In contrast to the coordinated efforts of U.S. East Coast mud bank feeders, the Cape Peron beach hunters rely on the incoming tide to avoid permanent stranding and are mostly solitary in their efforts. At Cape Peron, only calves of beach hunters have exhibited this behavior, and they are also at least a year or two old when attempting their first beach hunts. These observations suggest that the calves learn this behavior from their moms.

Two more accounts of foraging bottlenose dolphins illustrate the diversity and breadth of their hunting strategies, as well as the level of coordination and apparent communication involved with these activities. Dolphins generally swallow their fish whole, usually headfirst; it was thus believed that dolphins do not share fish. Where is the exception to this rule? In 2003, Sharlene Fedorowicz and her colleagues in Costa Rica saw two bottlenose dolphins (an adult male and an adult female) sharing a fish.[26] The female was accompanied by a calf, but her calf never touched the fish. Both adults were identified as part of the study population for a few years. The fish changed "hands" eleven times before it was consumed, and neither dolphin attempted to escape with the fish before the duet was complete. Although the function of this exchange is not yet fully understood, a minimum level of signal exchange between these two dolphins was required. And perhaps more—maybe the male was courting the female with a fish present? This behavior may represent a form of courtship display that we have yet to observe fully or understand. We hope one day we'll know the answer.

Off Cedar Key, Florida, Stephanie Gazda and her colleagues examined group hunting by two groups of bottlenose dolphins that exhibited role specialization. That is, these dolphins practiced a division of labor during coordinated foraging.[27] Individually identified dolphins assumed the role of "driver" while other dolphins in the group acted as a "barrier" to the fish. The driver chased fish to the barrier, and all took turns eating fish. The only other nonhuman animal for which a division of labor has been witnessed during foraging is the African lioness.[28] Females coordinate their hunts with center and peripheral roles. Success during coordinated hunting is higher when each female occupies a specific role, as compared with solitary hunting strategies. But how does each dolphin (or lioness for that matter) select his or her specific role? Are the roles assigned by hierarchy or skill? Also, what signals do they use to coordinate their foraging bouts? These questions remain unanswered to human observers, but one thing is certain: these individuals must communicate intent and action with their group mates if they are to engage in a division of labor for coordinated hunting.

The social life of various species and populations of dolphins bears various similarities with other animal groups. The family Canidae comprises roughly thirty-seven species, including African wild dogs, which hunt cooperatively, as well as other canids that share food and provide care for sick pack members and dependent young.[29]

Most dolphins live in a fission-fusion society. What does this mean? A fission-fusion social structure is characterized by individuals who associate in small groups that change in composition and size hourly, daily, seasonally. Bottlenose dolphins are a classic example of the variety in social structure found among delphinids. Researchers have ongoing, active projects studying bottlenose dolphins intensively in Sarasota Bay, Florida; in Monkey Mia, Shark Bay, Western Australia; along the Texas and California coastlines; in Moray Firth, Scotland; in the Black Sea; in The Bahamas; and in the waters of Japan, to name only a few locations.[30] Some dolphin communities, such as the populations studied in Sarasota Bay and Shark Bay, seem to be matrilineal, consisting of females and their accumulated offspring or sisters and other females. Calves within these groups will often develop stable relationships with one another over a period of years.[31]

The strongest bond between any two individuals for bottlenose dolphins has been confirmed as the association between mothers and their calves. Maternal investment ranges from two to four years in most studied bottlenose dolphins: the calf spends all of its time in the first year of life with mom, and occasionally with her adult female companions (often older female siblings of the calf or female siblings of the mother). Time with mom represents a period of learning and development. Calves are born with the ability to vocalize, to echolocate, and to whistle.[32] Calves must, however, still learn when and for what reasons it is appropriate to use the signals in their repertoire. They can practice these signals in the relative safety of the company of their moms in their natal group. In these situations dolphins often share babysitting or "calf-sitting" duties, in which some adult mothers watch several calves while the others forage in deeper water. This environment offers a setting for calves and other juvenile dolphins to learn from one another as well as to be directed by adults in their use of vocal, behavioral, and tactile signals. Stan Kuczaj and his students at the University of Southern Mississippi provided evidence that dolphin calves learn signals or innovate new behaviors more quickly in the presence of peers, specifically other calves of similar age.[33] The adult babysitters seem to oversee the activity of the calves and may intercede if a wrong signal is used, but association and interaction with similarly aged individuals, whether dolphins or humans, seems to offer a productive learning environment. The youngsters get to practice the signals that will form the foundation for all their communicative exchanges and to form social bonds that will endure a lifetime.

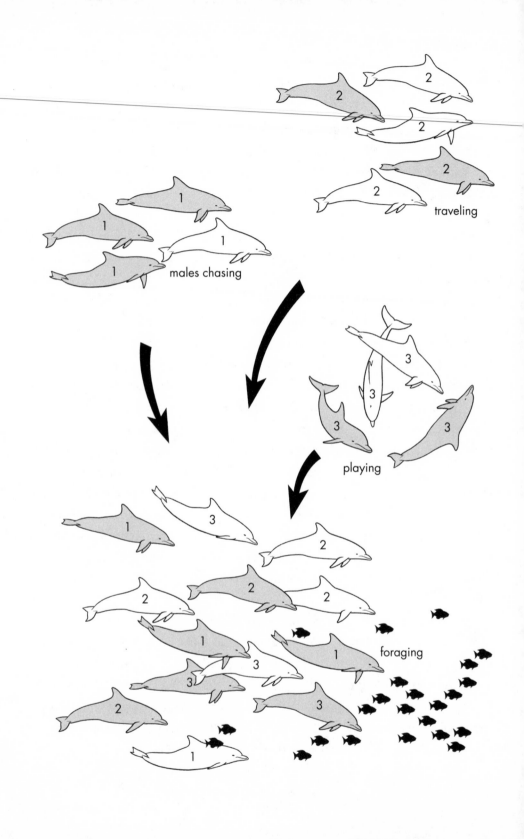

traveling

males chasing

playing

foraging

In the bottlenose dolphin social groups in Sarasota Bay and Shark Bay, subadult males usually leave their natal groups and form "bachelor" groups that often remain together indefinitely.[34] In these two study populations, sexually mature males often form partnerships or coalitions with other males and move between the female groups primarily for the purposes of reproduction. The level of association in these male friendships rivals the strength observed in the bonds between mothers and their calves.

Killer whales are the biggest species in the family Delphinidae. Northeast Pacific resident and transient orcas are sympatric populations—that is, diverse foraging and social strategies have resulted in separate stocks.[35] As a species, killer whales have a diverse diet, a strong matriarchal social structure, and acoustic behavior that includes dialects and other vocal cues that can be used to identify family pods.[36] (A

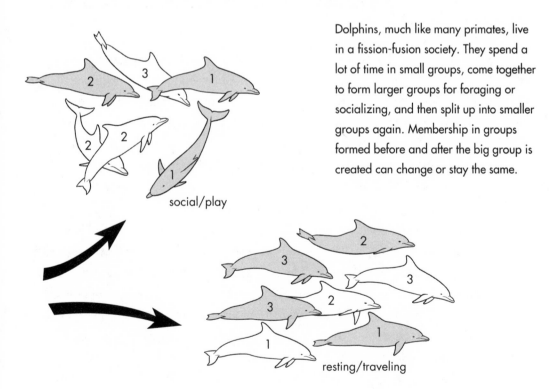

social/play

Dolphins, much like many primates, live in a fission-fusion society. They spend a lot of time in small groups, come together to form larger groups for foraging or socializing, and then split up into smaller groups again. Membership in groups formed before and after the big group is created can change or stay the same.

resting/traveling

Dolphins are exceptionally tactile and are often found in physical contact with one another. Calves especially stay close to mom as they develop in their first two to three years of life.

pod is a group of genetically related individuals. Thus, even though you often hear about dolphin pods in movies, on television, and in books, only orcas actually live in pods. Other dolphins are found in schools or, more generically, in groups.)

Killer whale society and maybe even orcas' relatedness are defined better by their vocal dialects than by their social behavior or travel associations. In this way, orcas are one of the most elegant examples of the importance of communication as a key to social structure and survival. The significance of vocal repertoire and dialect for society and culture is most easily observed in the highly stable "resident" orca groups of the northeastern Pacific.[37] The basic level of their social hierarchy is the matrilineal group, which typically persists over the animals' lifetimes—calves remain with their moms for life. The bonds among individuals of these matrilineal groups are so strong that it is unusual to see members of a pod apart from the group for more than a day. With the exception of some human populations, killer whales are the only mammals in which both genders remain with their maternal group for life. African elephant and lion females often remain with their matrilineal groups, much like many dolphin species, but the males leave the group at least occasionally, if not permanently.[38]

Within a matriline, the most stable level of social organization is the pod, consisting of one or more females and their offspring. Members of each resident pod share the same call repertoire and only occasionally share calls with other pods. The next level of cohesion is the clan: clans contain pods with similar vocal dialects, and it is believed that each clan originated from one pod.

Like other dolphins, killer whales produce three forms of vocalizations: clicks, whistles, and pulsed calls. In resident orcas, the vocalizations, or call types, within matrilineal groups are distinctive dialects. The calls of individuals within these groups can also be identified: each member has the same "accent," so to speak.[39] (Think of the distinctive accents of people from Boston, New York, Alabama, and Texas.) Dialects have also been confirmed in killer whales from northern Norway, and as more data are gathered, dialects will probably be found in other orca groups.[40] These dialects provide information on individual and pod identity not only to other

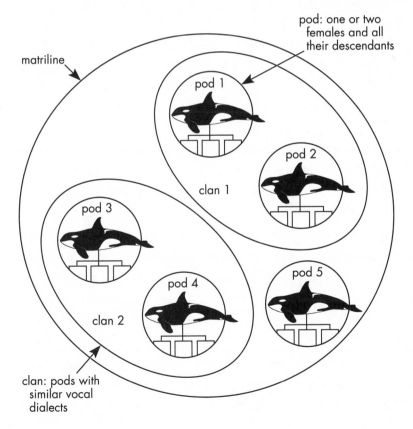

pod: one or two
females and all
their descendants

matriline

pod 1

pod 2

clan 1

pod 3

pod 4

pod 5

clan 2

clan: pods with
similar vocal
dialects

Orcas form a matriarchal society, with all offspring remaining with mom for life. The pod contains one or two adult females and all their descendants. A clan represents pods with similar vocal dialects. The killer whale pods in the Pacific Northwest have been studied for several decades, and some researchers can recognize the pods by their dialects.

orcas but to human observers. Recognizing individuals from your family group
by voice is important if you live in water with limited visibility, as do the resident
Pacific Northwest killer whales. The overlapping pods of this region of the Pacific
sometimes form superpods for mating and other social activities. The varied dialects
allow orcas to identify family and nonfamily members during these large social
gatherings. The dialects also help orcas in collaborative hunting: whistles help to
coordinate activity, whereas echolocation is used primarily to locate prey.

Resident killer whales rely heavily on their different calls to share information
when engaged in a variety of behavioral contexts. In the Pacific Northwest, the
transient population of killer whales has a range that extends farther offshore and
much farther south than resident pods. These transients live in smaller groups and
partake in a very different set of foraging tactics.[41] Vocalizations, as with other dolphins, are important to transient orca communication, but differences exist. Transients have adapted to more silent hunting practices and developed other methods
for information sharing while pursuing their prey. This makes sense—they are
hunting mammals with similar hearing and vocal ranges. Using echolocation or
other vocal cues to stalk their prey would be akin to announcing their intent and
presence: hunger would likely follow. As separate subspecies, resident and transient
orcas have each learned to adapt and maximize the resources of their niches.

When killer whale pods join, they engage in vocal and behavioral greeting ceremonies. African elephants have similar practices. When feeding, individual elephants
will range over large distances. When they later prepare to regroup, individuals will
signal vocally with a rumble and a special posture.[42] When in visual range but still
at a distance from group members, individuals may raise their heads, lift and spread
their ears, tuck their chins, and then rumble loudly while flapping their ears. Individual elephants or groups meeting may first raise their trunks in the air to smell or
continue rumbling, head lifting, and ear spreading. They may press heads together
and gently click or even intertwine tusks, sometimes while rumbling. This tusk
entwining varies from tusk-to-tusk battle observed when males, especially those in
musth, are competing.[43] I (Toni) have observed such elaborate greeting ceremonies
firsthand in two very different environments: orcas in the waters near my home in
the Pacific Northwest and elephants in Africa. Some of the elephant sounds I heard

during a greeting sounded as if the earth was purring like a giant cat. One of the things that struck me the most was that I could feel some of their low-frequency sounds. I felt what prompted Katy Payne to investigate their sounds. I can't imagine what people thought when feeling these sounds without knowing they came from elephants. Perhaps they thought there was an earthquake?

Although the greeting ceremonies among killer whales and elephants represent more vocal and postural salutations between individuals and groups, touch is used as a greeting in many species, as discussed previously. I (Kathleen) have observed flipper-to-flipper rubs in spotted and bottlenose dolphins used as greetings between individuals that have been apart in time or space, as well as possibly to greet human swimmers who have been absent for a few seasons. I believe that dolphins readily recognize peers, and maybe even individuals from other species, though data on the latter are currently lacking. I have seen adult females both in The Bahamas and around Mikura Island swim toward one another without a sound to exchange pectoral fin–to–pectoral fin rubs. This greeting always reminds me of the handshakes or hugs exchanged by people outside the baggage claim area in airports. Touch is a great reassurance, a reminder of a friendly relationship between individuals, no matter the species.

The method scientists use to identify distinct whistles is spectrographic analysis. A spectrogram is a picture of the sound we hear. This picture is actually a graph of sound

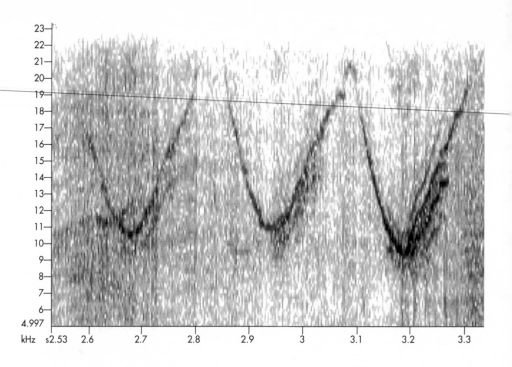

```
23–
22–
21–
20–
19=
18–
17–
16–
15–
14–
13–
12–
11–
10–
9–
8–
7–
6–
4.997
kHz  s2.53  2.6      2.7     2.8     2.9      3      3.1     3.2     3.3
```

Dolphins produce two broad categories of sound:
pulsed or click sounds and whistles. This spectrogram
is a visual picture of a dolphin whistle. Time is on the
x-axis and frequency is on the y-axis. This whistle has
harmonics and several inflection points or loops and
was produced by an Atlantic spotted dolphin in
The Bahamas.

frequency versus time. The plot of each whistle forms a line that is either straight or
curved. Whistles are narrow-band, frequency-modulated sounds that usually last from
a tenth of a second to three seconds with fundamental frequencies that range from 5
to 20 kHz.[44] A whistle's energy concentrates in a thin band on a spectrogram, forming
a distinctive pattern called the whistle contour. This contour is what has been described
as *distinctly characteristic* for specific dolphin whistles. In the scientific community,
there is much controversy about the nature and function of dolphin whistles.

In the 1950s, Carl Essapian and David and Melba Caldwell made the first recordings of dolphin whistles, which led them to suggest that individual bottlenose dolphins might produce distinctive whistles.[45] The Caldwells termed these

calls "signature whistles," which, by definition, represent 90 to 100 percent of a dolphin's vocal repertoire and were easily discernible from the signature whistles of other dolphins.[46] At about the same time, with different animals, William Dreher and William Evans reported that dolphins shared a large repertoire of signals, with particular whistles used in specific situations or to convey different emotional states.[47] Intriguingly, the production of some whistle types was attributed to all four species examined by Dreher and Evans, whereas other whistle types were produced exclusively by a single species. This is the crux of the scientific debate over signature whistles: Is a dolphin's vocal repertoire made of fewer specific and individually distinct whistles or of a larger set of more varied whistles? Two competing hypotheses feed this debate.

The *signature whistle hypothesis* postulates that signature whistles remain distinguishable and stable over years for each individual.[48] This hypothesis also stipulates that variation in some acoustic features in the whistle, such as duration, intensity, and number of whistles produced per minute, would change in different behavioral contexts and so provide information other than individual identity. Still, the main information conveyed by each whistle is individual identity. Spectrographic analysis is used almost universally in the study of animal sounds: elephants, birds, whales, dolphins, and even humans. Dolphins produce additional whistles in their repertoire that are not signatures: these whistles appear in low numbers and have been called "aberrant."

In contrast, the *whistle repertoire hypothesis* stipulates that dolphins share a large repertoire of whistles with other individuals. This hypothesis advocates that information concerning individual recognition of an animal could be represented in subtle differences in the acoustic features of each whistle contour, similar to the voice cues found in primates. Researchers Brenda McCowan and Diana Reiss argue that by calling whistles that are different from signature whistles "aberrant" or variant, previous studies may have suppressed the function or position of these whistles within the dolphin vocal repertoire.[49]

Both hypotheses are based on results from work conducted on captive dolphins, though both also have support from observations of dolphins in the wild. In Monkey Mia, vocalizations exchanged within mother-calf pairs were recorded when they were separated.[50] The calves seldom produced whistles when in contact with their mothers but whistled often when apart from mom. The whistles were individually distinct. In a study in Sarasota conducted for more than fourteen years, the vocal exchanges between mothers and their calves were recorded. One major difference between this study and the one conducted in Australia is that the dolphins in Florida are temporarily held with nets in shallow water.[51] During the time span of the study, the signature whistles of individual dolphins did not change. Even calves, recorded as young as one or two years old, used stable signature whistles. Although these studies discovered important aspects in the communicative aspects of whistles, they share a major drawback. They did not record dolphins during complex social interactions. Instead, the calves produced whistles while leaving or returning to their mothers or in isolation from peers. Dolphin society is an intricate, fluid assembly of fission-fusion interactions. To record a complete repertoire of dolphin whistles, one would need to record them in a variety of social contexts. In all animal communication systems, individuals are more likely to convey messages, to share information or probe for it, when in the presence of other individuals. Therefore, these studies likely presented a small part of these dolphins' normal whistle repertoires.

The vocal data from spotted and bottlenose dolphins that I (Kathleen) have collected over the years at first led me to reject the signature whistle hypothesis. With my recording system, I am able to match sounds to individual dolphins and their actions for about 40 percent of all my data. At all of my field sites, I have never had the same number of distinct vocalizations as there are number of dolphins. And besides, why would dolphins, accepted as highly evolved and intelligent beings, swim around all the time whistling only their name? The data I gathered, from socially interacting groups from underwater, with dolphins almost always in visual range of peers, showed no support for individually distinct whistles per identified dolphin. That is, I recorded about ten categories of whistle contour patterns that several dolphins produced. It would be safe to say that I was solidly in support of the whistle repertoire hypothesis for about a dozen years. Then I began studying dolphins in captivity along with my work on wild populations. I chatted with trainers about the distinct whistles emitted by specific dolphins that the trainers readily recognized. I heard these whistles myself. I began to rethink my archive of data and my interpretation in light of my recent observations. I now believe the hypothesis proposed by Vincent Janik and Peter Slater, which suggests that individually distinct whistles are likely used to maintain group cohesion or contact among individuals separated by distance.[52] That is, dolphins out of sight of friends or group members use a specific whistle (defined by us with a spectrographic contour pattern) to remind their buddies that they are nearby or that they want to regroup. My data then made sense: signatures or individually distinct whistles would be hugely underrepresented in my data, since the dolphins that I observe are always in view of their peers.

Reciprocally altruistic behaviors between nonrelated animals require that individuals be able to recognize one another individually. This is particularly important when all animals within a given group are not genetically related. The ability to recognize individuals in large social groups facilitates cooperative alliances and other associations.[53] Individually distinct whistles could be very useful for maintaining individual bonds and social hierarchies within the fission-fusion society characteristic of many dolphin species.

Vocal learning—the ability to modify one's vocalizations in response to auditory experience—has been confirmed in only a few mammal species, for example, humans, chimps, elephants, and dolphins.[54] A flexible and open communication system is enabled by vocal learning; animals may also learn to imitate signals that are not species typical.[55] Dolphins can imitate a trainer's whistle, and an African elephant demonstrated an ability to imitate truck noises.[56] Vocal learning is an evolutionary response to maintaining relationships in a fluid fission-fusion society. Learning to imitate peers is a communication tool for the maintenance of individual-specific bonds. In many instances, vocal learning is also considered evidence of higher cognitive processing.

mocking Birds imitate other birds + other sounds such as cell phones ringing.

For many species, nonvocal communication is an important component of information exchange. Visual displays or signals include body postures, coloration patterns, and elaborate sequences of behaviors—anything that can be seen. Visual signals can indicate the species or individual identity, age, sex, or reproductive status. Visual cues such as postures or behaviors are likely to signal intent and demeanor of a sender, as well as provide insight to the meaning concluded by a receiver. Land mammals exchange much information through subtle or overt kinesic (gesture or movement) expressions, including changes in facial expression (a dog's

A group of spotted dolphins is tightly formed and circled against a single bottlenose dolphin who tried to infiltrate the group . . . possibly for mating or maybe for play.

snarl), irregularities in respiration, and overall body movement. When fighting, many animals have the ability to make themselves seem larger. Elephants will flare out and flap their ears when agitated.[57] Bears rear up when attacking or defending their young. Birds fluff out their feathers and cats fluff their fur. Dolphins flare their pectoral fins while producing bubbles during a fight. In 1994, I (Kathleen) observed a group of spotted dolphins fighting in small gangs: groups of three to four individuals were going head to head, vocalizing loudly, ramming and hitting one another, and producing large amounts of bubbles (streams, trails, and clouds). I have no idea what they were fighting about. The dolphins that were vertical and head up below another group produced more bubbles and flared their pectoral fins more than the dolphins above them. Based on this scene and many other observations, I believe dolphins engage in the art of intimidation. If you can bluff another individual away with bluster or bubbles and flipper flaring, then you might avoid

The spotted dolphins are successful in fending off the bottlenose dolphin in this instance.

the more energy-consuming actions of a true fight. And you stand a better chance of not getting injured!

Dolphins also use synchronous behavior to signal to peers and opponents, depending on the context. Some male bottlenose dolphins form friendships that last a lifetime; they are rarely, if ever, seen apart.[58] These male pairs may work with other male pairs to herd females or compete for access to female dolphins for mating. Within each pair, the male dolphins often engage in highly synchronous actions, often surfacing to breathe, leaping, diving deep, or traveling in perfect unison. This synchronous behavior is a form of communication, not only to each member of the pair but also to potential opponents or even to the pursued female. The males are using this nonvocal behavior to illustrate their bond and camaraderie. Their synchrony showcases their ability to coordinate closely and work together toward a goal. Competing male dolphins, as well as the female being pursued, know that the synchronous pair is a solid team. In this way, by their harmonized movements, the male dolphin pair is honestly signaling their intent, either toward the female or toward the competition.

Communication between individuals is not always honest and forthright, and the transmission of misleading information is used to deceive and manipulate the behavior of others. Deception is observed in a variety of land animals, including vervet monkeys, great tits, and chimpanzees.[59] Human beings fib, fabricate, and tell lies to further their cause. Think of the sweetly smiling boy who asks for dessert without having finished his peas, hiding them under the last scoop of mashed potatoes. Why else would we learn at a young age to lie about the source of the spilt milk or blame the dog for the dirt tracks on the living room carpet? Given the highly developed communication systems of dolphins, it is likely that they also employ deceptive signals. Because of the relative lack of detailed studies on dolphin communication, as compared with terrestrial mammals, however, it is not surprising that deception among dolphins has not yet been conclusively demonstrated.

Evolution may have contributed in a variety of ways to geographic variation in whistle characteristics within and among dolphin species. Acoustic variation in habitats affecting sound transmission properties may influence the vocal qualities of some dolphins. Over time, habitat differences can affect the genes of populations or species so that they are able to produce different types of vocalizations. The distinct dialect of each resident Pacific Northwest killer whale pod cannot be attributed to reproductive isolation, since these pods not only share much of the same habitat but also socialize frequently. The impact of geographic distribution and cultural selection has also been explored in the freshwater and marine subspecies of the tuxuci dolphins.[60] In this case, some individuals within the two subspecies inhabit different ecological niches, whereas others overlap. In this and other dolphin species, habitat and genetic differences may always be an influence, but as in the case of the resident orcas of the Pacific Northwest, variation in vocalization, as well as other behaviors, are more likely attributable to social learning and cultural transmission across generations.

Chapter 3 Eavesdropping

When a wild creature comes close because it chooses to, we can see more clearly into its world; a world unclouded by fear is more transparent, more easily understood.

MARK AND DELIA OWENS, *Secrets of the Savanna*

Late one balmy July afternoon aboard ship in The Bahamas, we saw a very active group of dolphins. They were leaping and splashing and rolling all over each other in the distance, and they were headed our way. We were here to videotape them underwater . . . so we jumped in and waited for them to arrive. First our ears were assaulted by a loud and disparate "parade" of harsh sounds: squawks, intense whines, and shrill whistles. Within seconds, more than thirty dolphins zoomed toward us, twisting and curling around one another—torpedoes with minds of their own! They stopped abruptly, within an arm's length in front of us. With their jaws clapping, they separated and faced off from one another, head to head. Then, as pairs and triplets, like underwater street gangs, these dolphins tail kicked, rammed, charged at, and bit one another: the ensuing mêlée seemed a confusing jumble. The vast array of vocalizations continued intensely. Adrenaline was most certainly running high, and not just in the dolphins! Yet just as swiftly as they had darted into view, the fighting stopped. The dolphins simultaneously became almost silent and, within their little gangs, began caressing and rubbing one another. Without exception, all the members of these little groups were placing their flippers on each other's sides. And then, almost as abruptly as it ceased, the "fighting" resumed about two minutes later. This alternating pattern of fight and caress, fight and caress continued for the fifteen minutes they were in our view. In this context, it seemed as if the dolphins' gentle touch and pectoral fin placement to one another's side was reaffirming their friendship and fighting allegiance.—Kathleen

---

Listening to dolphins vocalize underwater is like being submerged within a highly orchestrated, yet flowing cacophony of sound. Fluid, high-pitched whistles seem to explode and then evaporate on the current. Abrupt chirps and squeaks, along with staccato squawks and creaking-door clicks, cause your eyebrows to rise and make you chuckle (which can leak water into your underwater mask if you're not careful). When a dolphin turns toward you and explores you acoustically with echolocation, you may even feel the buzz of sound waves in your bones and muscles, as you reveal more detailed aspects of your physiology to the dolphin by sound waves: more information than the eyes can see. While vocalizing, dolphins may be motionless or may coordinate their sounds with a plethora of movements ranging from subtle postures to dramatic aerial leaps. Despite the sometimes-dramatic differences between the senses of dolphins and humans, eavesdropping on dolphins gives us a window through which we can imagine what their communication is really like.

When we ordinarily think of eavesdropping, we usually think of someone listening to a conversation without being observed—from behind a closed door, around a corner, or on a telephone extension. Remaining undetected allows us to learn much information from, and about, others. But snooping on dolphins is a whole different story than eavesdropping on a sister's phone conversation. Eavesdropping on another species is not always easy, especially when dolphins can usually detect us well before we detect them. Furthermore, dolphins live a graceful, mobile existence in a fluid world in which we are—at best—capable but relatively clumsy. Fortunately, we can eavesdrop on dolphins even when our presence is known simply by establishing ourselves as discreet, noninvasive, predictable, and—frankly—boring. In this sense, dolphins learn to habituate, acclimate, or at least tolerate human presence. Eavesdropping is one of the best ways for us to learn more about dolphin communication and social life.

Since the early 1960s, many scientists have pursued the systematic study of communication in animals. A researcher's ability to eavesdrop on his or her subjects has become an invaluable skill. The most famous researcher in this regard is Jane Goodall, who pioneered scientific eavesdropping of another species by gaining the tolerance, trust, and acceptance of a group of chimpanzees in Tanzania in the 1960s.[1] Katy Payne discovered elephant infrasonic (frequencies below human audible sound) communication literally by feel.[2] Before Payne and her colleagues made this discovery, no one knew how, or if, elephants coordinated and met up with one another over long distances. Now researchers can record these infrasonic elephant conversations using sophisticated microphones without even being near their subjects—the ultimate in eavesdropping.

As much as we try to be discreet dolphin voyeurs, we must remember not to become overzealous paparazzi in our quest to learn more about them. It is sometimes hard not to feel like a well-intentioned stalker with video or still camera (sometimes both) in hand while in the water with dolphins. This situation, however, is not nearly as surprising as finding oneself surrounded by the same dolphins we are trying to observe unobtrusively.

Sometimes curious and playful, possibly lonely, and even mischievous dolphins are not always content with researchers simply being subdued passive observers, which presents both challenges and opportunities. As was told in the days of Greek myth, dolphins are unique among wild animals in that individuals will from time to time approach humans socially, even without being offered food. Interacting with dolphins can at times teach us about how they communicate with one another, which we detail later.

People have been interested in communication between nonhuman animals for centuries for more than scientific purposes. By recognizing alarm calls used by, for example, vervet monkeys to alert one another to the presence of predators, indigenous peoples learned to avoid shared predators. Familiarity with the behavior of potential predators also helped humans avoid becoming a menu item. For example, becoming more aware of the signals that domestic dogs exchange can be helpful in detecting the more subtle indications of a dog alerted to an intruder (before the dog barks) and of a dog intending to bite you (even before the dog growls).

A dolphin looks so graceful that it is hard to imagine that its body has been shaped for life in water rather than beauty.[3] The aquatic environment has extensively sculpted how dolphins communicate with one another. As we discussed in chapters 1 and 2, they have an impressive variety of signals and a remarkable range of hearing and sound production capabilities. These abilities seem to have been adapted to each species' habitat. River dolphins often live in murky waters and are known to have relatively poor eyesight but exceptional echolocation aptitude. It is unlikely that river dolphins would rely on visual communication as much as the use of hearing and sound production, although this hypothesis remains to be tested. Conversely, Atlantic spotted dolphins in The Bahamas usually swim through water with visibility at least ninety feet (about thirty meters) in depth; Kathleen's studies show that they produce only about a third the number of whistles as the Indo-Pacific bottlenose dolphins found around Mikura Island, where visibility is generally not as clear underwater. Thus, at distances where they can see one another, dolphins seem to rely more heavily on visual, behavioral, and postural cues than on auditory cues to exchange information.

Postures or gestures are distinct visual displays and are useful for close-range communication among dolphins. Visual signals provide an alternative to or work in synergy with acoustic signaling. Coloration and other physical characteristics of the dolphin body can modify and expand how they share information, as well as what specifically they might be trying to convey to peers. Body coloration, body size, fin shape, and other physical features can reveal age, gender, reproductive status, and species or be a component of communicative signaling. Although the length and shape of their limbs have been modified for swimming, dolphins are adept at using subtle shifts in posture or slight body movements to convey information in their three-dimensional world. For instance, aggressive threats in dolphins are often expressed through a direct horizontal approach that can be coupled with jaw claps, head shakes, body hits or slams, or the emission of a bubble cloud.

During fights, dolphins may orient themselves head to head and flare out both flippers in an exaggerated way. This makes the dolphin look bigger and may work to scare off an opponent.[4] The group of fighting spotted dolphins described above

Even though she refuses to set foot on a boat or to swim in the sea, Umi the mighty sea beagle has guided Kathleen many times as she reviews whistle data from the dolphins around Mikura Island, Japan.

featured many types of flipper use: they were flared, used in affectionate contact between "gang" members, and guided their swiftly moving bodies as each dolphin careened around the others while delivering swift tail kicks to opponents. For most dolphin groups, when fights are observed, the gangs are actually pairs or triplets of males vying for access to a female for mating. This group of spotted dolphins, however, has defied complete classification because juvenile, subadult, and adult males as well as females were involved. Maybe they were irritated over some other aspect of social life . . . Could they just have been ornery? Or was there a disagreement over a foraging site? Additional observations of similar activity in other dolphins or watching this video for more hours might shed more light. Still, this fight and caress sequence illustrates how one action can have different meanings depending on the context or ensuing activity of the individuals as a whole. Flipper flaring by dolphins seems to function similarly to ear flapping by elephants.[5] Flaring out one flipper (or one elephant ear) during nonaggressive or nonthreatening interactions sends a very different message. Perhaps the flare signifies affection, as compared to the increased size that the double-pectoral flare is likely to convey.[6] When a dolphin swims, the flippers give the dolphin a more refined directional control. A slight shift in the angle held by a dolphin's flipper might signify a change in swimming direction to others in the group. Thus, the similar action of the flipper used in a different situation may send a different message to receivers.

Have you ever felt as if you were going around in circles? Dolphins must share that feeling. At least, dolphins often seem to circle people swimming near them. We have both witnessed these acrobatic spinning sessions. A typical spinning scenario usually involves a single, relatively young dolphin swimming in increasingly tighter circles around a person who awkwardly spins around with the dolphin. In this way, we have found ourselves becoming our own "subjects." Although this circling behavior is not uncommon among dolphins, especially during play, it seems more common during interactions with swimmers. Perhaps this is because humans are simply less agile in the water, which makes us prime targets for amusement by dolphins in need of cheap entertainment.

Conversely, perhaps this circling behavior manifests differently when it occurs with a dolphin partner. We imagine the ballet between two spinning dolphins to be like a pas-de-deux in *Swan Lake*, as opposed to a George Carlin show, which could characterize the human-dolphin spinning pair. Still, we have a sneaking suspicion that there is a bit of mischief on the part of the dolphins who engage in circle swimming—especially since juvenile dolphins are so frequently the perpetrators. Or could the circling behavior be an attempt to communicate something visually to

From their blowhole, dolphins blow a variety of bubbles ranging in size from thin streams of tiny bubbles to trails of larger, more oddly shaped bubbles to the largest bubble clouds, produced when dolphins are fighting. The function of these bubble emissions is only recently being better understood.

peers or maybe to the humans involved? We have tried to analyze this behavior from videotape rigorously—counting how many times it happens and even attempting to find meaning in whether the dolphins circle clockwise or counterclockwise. In one study, Toni found that these "circle the human" bouts occurred 7 percent of the time that spotted dolphins in The Bahamas accompanied human swimmers—but the meaning of which direction they circled was not apparent (perhaps it is a matter of left- or right-handedness?).[7] Clearly, the dolphins have had us "going in circles" out of the water as well as in it.

There are certainly other times when the observer becomes the participant. In an episode that could easily be called "who's watching who?" I (Toni) fondly remember an afternoon with other people on a boat trip in The Bahamas when one male passenger posed as Poseidon with a boat pole as his trident and a handcrafted seaweed crown on his head. We didn't know that a spotted dolphin was lurking below the water surface watching us. After several minutes, the seaweed crown was inadvertently tossed into the sea. And soon we saw the dolphin wearing the crown in a parody of sorts of our behavior.

Visual signals often become tactile signals, especially in cases where underwater visibility is limited. Dolphins are exceptionally tactile; where and how dolphins are touched conveys different levels of information, or signal content, to their peers. Extensive contact and rubbing occur in both captive and wild dolphins during play, sexual and social activity, between mothers and their calves, and among juveniles. Dolphins use their rostrum, pectoral fins (flippers), dorsal fin, flukes (tail), belly, and even the entire body when touching one another.

I (Kathleen) am particularly fascinated with the exchange of touch between dolphins and, for several years, have been investigating how dolphins might use flipper contact (touches and rubs) to communicate on different topics with peers. I have collected data on wild Atlantic spotted dolphins in The Bahamas, wild Indo-Pacific bottlenose dolphins around Mikura Island, and captive bottlenose dolphins at the Roatan Institute for Marine Sciences (RIMS) in Honduras. While swimming among dolphins and applying the same protocol for observations of both captive and wild dolphins, I use a mobile video/acoustic system to gather behavioral and audio

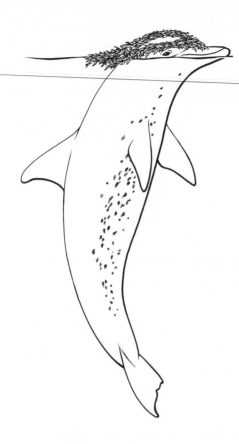

A subadult spotted dolphin delighted in the discarded seaweed wreathe that humans had been playing with aboard ship just a few minutes before. The dolphin frolicked and enjoyed her seaweed crown and seemed to imitate the human antics she had stealthily observed.

data. Where on the body, with whom, and during what context a touch or rub occurs significantly affects the meaning of a flipper contact exchanged between dolphins. An excited calf can be calmed by a mother's pectoral fin, but that same flipper can also discipline a more rambunctious youngster when needed.

In my study of flipper contact, I needed to create behavioral definitions so that other scientists could follow my procedures or replicate my methods, if needed. And referring to the "dolphin whose pec fin touched the second dolphin's body" was wordy and growing tiresome. So, in my new behavioral vocabulary, the *rubber* is the dolphin whose pectoral fin touched a second dolphin's body, while the *rubbee* is the dolphin whose body was touched by another dolphin's pectoral fin. (These terms have produced plenty of chuckles during scientific presentations!) Either the rubber or rubbee can initiate or receive flipper contact. As I collected data and observed how dolphins touched or rubbed one another with their pectoral fins, I began to see patterns in the data and subtle differences between the two wild dolphin groups. Older bottlenose dolphins at Mikura Island shared more pectoral fin contact, whereas younger spotted dolphins in The Bahamas engaged in more flipper touches and rubs. In fact, dolphins from both groups seemed to choose their contact partners; same-age, same-sex pairs shared more rubs and touches than other pairings. In addition, the rubbee seemed to initiate more contact among the dolphins at Mikura Island, whereas the rubber engaged in more contact among spotted dolphins in The Bahamas. I'm still not sure about the age difference between both dolphin groups, but I think that the

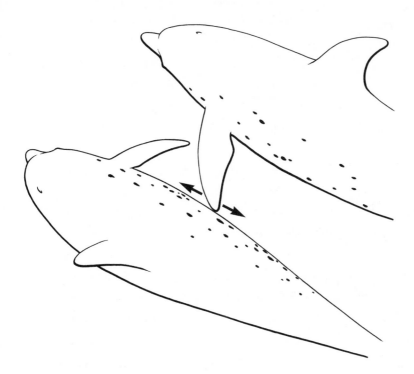

Dolphins often share pectoral fin rubs where the flipper contacts have a specific meaning. The dolphin rubbing her pectoral fin along the side of another dolphin between the flukes and dorsal fin is asking for something. That something depends on the ensuing activity but might be continued rubbing behavior or maybe even for help in a fight.

rubber-rubbee differences might be explained by environmental factors. Rubbees may solicit more contact from rubbers at Mikura Island because these dolphins do not rub against the rocky boulder bottom as much as the spotted dolphins rub on the sandy sea floor in The Bahamas. I remember that for every entry to swim with and observe spotted dolphins, there was at least one dolphin rubbing its body into the soft sand.

With these unexpected, subtle differences observed among two wild dolphin groups, I was intrigued to see what I would find when I began my studies of the captive dolphins at the Roatan Institute for Marine Sciences. Besides looking at how dolphins share pectoral fin contacts, I made "blind" observations of each dolphin trainer with each dolphin at RIMS. I was curious to see whether any patterns in behavior that I observed among the dolphins might be affected by their interaction with the trainers; if a dolphin did not use much pectoral fin contact with other dolphins, did that dolphin use more or less or the same amount with a person? The trainers did not know that I was occasionally documenting, or eavesdropping on, their contact with dolphins during training sessions. It turns out that the trends in behavior that I documented among the dolphins held true in their associations with their trainers. If a dolphin engaged in frequent tactile contact with other dolphins, that individual was also likely to solicit more touch from their trainer, and vice versa.

Somewhat surprisingly, I also found many more similarities in touches and rubs between the dolphins at RIMS and both wild dolphin study groups. This is the most fascinating part of behavioral science—just when you think you have the answer, an alternative explanation pops up for consideration.

Body rubs and other contact are also important: whether traveling, resting, social-izing, or playing, dolphins are often seen in physical contact with other dolphins. Contact between dolphins can be modified to increase the information content: who, where, and how animals touch, as well as the intensity of a touch, factor into a signal's meaning. Going back to the play-versus-fight scenarios illustrates this point. During play, hits, harsh sounds, bites, and other aggressive actions are modified by oblique angles of approach and frequent affectionate rubbing to indicate that the ongoing activity is playful and not seriously aggressive.

Dolphins often travel in contact with one another. They are very tactile and will rub each other's bodies as they swim through the water.

Dolphins leap for a variety of reasons. They spin, somersault, perform back and front flips, and more as they glide through the air. Dusky dolphins are known for three types of leaps, while spinner dolphins are named for their high spinning leaps.

When a dolphin leaps from and then reenters the water, its body hitting the sea surface creates a sound that can carry for several miles. This breaching or leaping behavior often indicates general excitement deriving from any of several causes, including sexual stimulation, location of food, a response to injury or irritation, or the need to remove parasites. Dusky dolphins are well known for three types of leaps they exhibit in association with three stages of cooperative feeding: headfirst reentry leaps, noisy leaps, and social, acrobatic leaps.[8] The last two create sounds that probably signal the state of activity to peers. Since these leaps occur nearer to the end of feeding, they may signal the shift from "dinnertime" to "dancing." Noisy leaps could also act as a sound barrier to disorient prey and keep them tightly schooled—a signal that is to the dolphin's benefit, if not to the fish. Spinner dolphins, on reentering water after a spin, a breach, a back slap or a head or tail slap, generate omnidirectional noise that travels over short distances. The leaps and spins of spinner dolphins seem designed to produce this noise, because these actions commonly occur at night, when visual contact is limited.[9]

When conveying a sense of threat or frustration, dolphins will tail slap the water dozens of times, creating loud, low-frequency underwater and aerial sounds.[10] Just as you would not approach a growling dog, it is best not to approach a tail-slapping dolphin. The same goes for dolphins exhibiting an "S-shaped" posture, which can be followed by aggressive behavior.[11] But both of us have witnessed S-postures and tail slaps used in nonaggressive contexts, too. For instance, juvenile spotted and spinner dolphins practice S-postures and tail slaps when playing, and dusky dolphins use tail slaps to corral fish for group feeding.[12] As we saw in chapter 2, the percussive sound of a jaw clap accompanied with a direct approach or an aggressive posture is also a threat or warning signal. The social functions of nonvocal acoustic dolphin signals appear to be limited mainly to long-range communication to regroup with peers or as expressions of excitement, annoyance, and aggression.

Dolphins often couple nonvocal acoustic cues with other behaviors to exaggerate a signal and its message. Sometimes two or more male bottlenose dolphins will herd or corral a female to mate with her.[13] Not only do the males create loud jaw pops, but they tail slap near the female, push and hit at her body, and approach her aggressively. A female pursued with these signals is unlikely to misunderstand her suitors' intent.

The emission of bubbles from the blowhole may serve more as a visual signal than an acoustic one. Some researchers use a bubble stream that occurs with a whistle as a cue to identify a vocalizing dolphin from within a group.[14] Unfortunately, no study has examined how dolphins might use bubble streams in relation to sounds. We do know that juvenile dolphins often emit bubble streams from the blowhole when they appear excited. When we are studying dolphins underwater, juveniles seem more interested than older animals in us. Young dolphins are inquisitive and often emit streams of bubbles while whistling. Could these bubble emissions be related to age—that is, are excited youngsters "leaky"?

Dolphins assume an S-posture as adults when they are fighting. Juvenile dolphins practice the S-posture when playing with other young dolphins. The rostrum is pointed up, the back is down, the peduncle is up, and the flukes are down, creating an "S" shape.

As we saw in chapter 2, dolphins produce all their vocal sounds by passing air back and forth between four sets of air sacs behind their melon. Maybe refined control of the air sacs, and thus bubble production, comes with age? Alternately, maybe the bubbles have a behavioral role. Dolphins produce many types of bubbles: bubble streams, bubble trails (slightly larger bubbles than streams), and bubble clouds. It is possible that bubbles sometimes function as a tool. Dolphin vocalizations can be powerful and painful: dolphins may use sound to injure the fish on which they prey.[15] They might do the same to one another. Bubble clouds may form a shield that deflects an opponent's aggressive, percussive sounds. The sounds are reflected off the air pocket created by each bubble cloud and away from the defending dolphin.

Bubbles likely have an array of functions depending on the activity and relations of the dolphins involved. We have both seen dolphins emit large underwater bubbles (befitting a cartoon caption) immediately after experiencing a novel or attractive sight or sound. Toni saw this often with dolphins who accompanied certain boats and swimmers. The dolphins sometimes emitted a bubble burst immediately after hearing a sound such as the splash made by someone jumping into the water or a boat engine revving. These bubble bursts happened so predictably with one gregarious and curious dolphin that we dubbed him the "cartoon character" dolphin.

Dolphins do not use their vocal cords to make sounds, yet they make an incredible variety of sounds that emanate from their head (see chapter 2 for details regarding dolphin sound production). These sounds are broadly divided into clicks, which include "pulse" sounds, and whistles. The sounds can be powerful, and dolphins are capable of fine control of the intensity of their vocalizations. All toothed whales make click (pulsed) sounds. High-frequency click trains constitute dolphins' echolocation or sonar, which they use to investigate their environment and locate food. Because not all dolphins whistle, the use of clicks or echolocation for communication cannot be ruled out. Other mammals also have the ability to communicate using pulsed sounds or clicks: bats are specialized to use echolocation in air and have evolved specialized ear and nose leaves to help focus, direct, and collect sounds.[16]

Click trains with high repetition rates are called burst-pulses. In burst-pulses, the

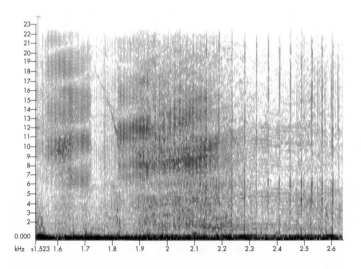

Echolocation clicks and squawks are shown on this spectrogram. Dolphins squawk when they play or are fighting.

clicks are emitted so quickly that to the human ear they resemble continuous sound rather than a series of clicks. Squawks, whines, barks, and moans are examples of these sounds. It seems that burst-pulses are used to communicate among individuals rather than to discern information about the environment. We have seen dolphins emit squawks, whines, and barks when playing and fighting.[17] These harsh sounds probably indicate excitement or frustration. The directional characteristics of many pulsed sounds, the relative ease with which they can be localized, their variability, and perhaps the power (that is, not only the intensity of the sound but the tactile effects of a loud, powerful sound) with which they can be produced enhance their value and usefulness as communication signals.

Another possibility for dolphins to exchange information via pulsed sounds may be through passive communication—that is, eavesdropping on other dolphins. This term is perhaps misleading; we are referring not to what people know as eavesdropping but rather to "echoic eavesdropping," which dolphins might use to share information. One individual might listen to a nearby dolphin's clicks and associated echoes toward an object rather than produce its own click train. After all, it takes less energy to listen than to click. Alternatively, dolphins might listen rather than produce clicks to avoid potential signal jamming: confusion might occur if all dolphins traveling together in a group were actively echolocating on a particular target. This suggests that eavesdropping might be something dolphins actively practice. We know that dolphins can listen to other dolphin echoes. Researchers at The Living Seas, Epcot Center, showed that a bottlenose dolphin trained to listen only to the clicks of an actively echolocating dolphin could discriminate targets.[18] This observation provides evidence that dolphins can detect and interpret the echoes of another dolphin's sonar. But we still do not definitely know if wild dolphins have these abilities.

These results raise an interesting question about the etiquette of echolocation use and how individual dolphins within a group know when it is proper to echolocate and when it is not. One of Kathleen's doctoral students, Justin Gregg, from Trinity College Dublin, Ireland, is investigating this topic in his research of the bottlenose dolphins at Mikura Island. Justin is using data collected with Kathleen's mobile video/acoustic system and echolocation click detector (more about these instruments later) to see whether these wild dolphins echoically eavesdrop on one another; he has created a three-dimensional method for measuring the angle between the heads of two dolphins as they appear on video, swimming next to each other and toward the mobile video/acoustic system.[19] As we learn more about dolphin social behavior and communication, we are able to answer research questions and refine them in order to direct future study. We are beginning to see that wild dolphins do indeed use their clicks for social reasons and not just for foraging or short-range navigation. Each nugget of information that we glean from our data and results gives us a better understanding of the many ways dolphins communicate, as well as how these methods might be used alone or in concert.

Many dolphin whistles are within our hearing range of about 2 kHz to 20 kHz and last from milliseconds to a few seconds. These sounds can have a rich harmonic content (think of a symphony) that extends into the ultrasonic range of frequencies—three to four times higher than the range of human hearing. Musicians refer to the perceived whistle frequency as the *pitch*; the *duration* is the length the whistle occurs. Graphing these two bits of information provides the picture, or contour pattern, of the whistle on a spectrogram. When we examine these whistles as spectrograms, we see that they vary greatly in contour and shape, representing changes in pitch over time from simple up or down sweeps to warbles, U-loops, and inverted U-loops. Whistles with harmonics sound more "full" to the human ear, and perhaps to the dolphin ear, and are represented by multiple lines above the main whistle contour. Conversely, clicks on a spectrogram are represented more by vertical parallel lines. The main reason for this picture difference is how they are modulated: whistles are frequency-modulated pure tones, whereas clicks are amplitude-modulated and express energy across multiple frequencies. Click or pulsed sounds with high repetition rates such as burst-pulse whines or squawks can resemble whistle contour patterns on a spectrogram. Burst-pulses consist of hundreds, even thousands, of clicks per second, and the resulting contour can appear blended. The spectrogram is the best method for visual representation of dolphin sounds that we study.

Whistles function for communication, but as we have discussed, not all dolphin species whistle. A few species, such as Hector's dolphin and Commerson's dolphin, do not. We do not know why some whistle and others don't. The characteristic of whistling seems to appear in species with specific ecological or social conditions, though this assertion is not all-inclusive. For instance, all whistling species are highly gregarious. Most nonwhistling species share a characteristic of low gregariousness, but Hector's dolphins live in very social groups and do not whistle.[20]

Because whistles have a relatively lower frequency than pulsed sounds, they travel farther in water. Low-frequency sounds have a longer wavelength than higher-pitched sounds, which allows them to travel farther—their wave shapes can move around objects rather than get blocked or bounced back by these obstacles. Think of a length of rope: two people hold the ends and move their ends slowly up and down. This would create long wavelengths. Conversely, if they moved their arms up and down swiftly, they would create short wavelengths. The rope and waves represent the sounds that we hear: long versus short and the continuum between these extremes.

For further illustration, consider these nonaquatic examples: elephants and birds. Elephants, as we have seen, use low-frequency sounds well below human hearing to communicate over long distances. Atmospheric conditions certainly affect sound transmission, generally speaking, low frequency equals longer wavelength, which equals greater range over which to communicate. Most birds, in contrast, produce high-pitched sounds. Think of the bird songs you hear each morning and note how their songs do not carry very far because of their high frequency.

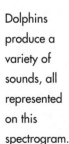

Dolphins produce a variety of sounds, all represented on this spectrogram.

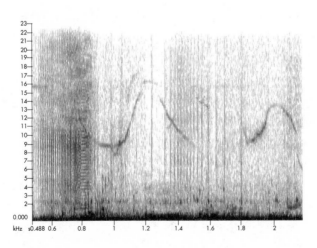

Dolphins that whistle can produce whistles and clicks simultaneously. Whistles might provide a potential vehicle for maintaining contact and coordination among group members while searching for food with echolocation. Whistles and sonar do not overlap in fundamental frequency, thus minimizing potential masking effects or miscommunication. Species, regional, or individual specificity in whistles facilitates identification of schoolmates or familiar associates, aids in the assembly of dispersed animals, and helps maintain coordination, spacing, and movements of individuals in rapidly swimming, communally foraging herds.

A wealth of data has been gathered on dolphin vocalizations and their concurrent nonvocal behavior. Hawaiian spinner dolphins, bottlenose dolphins, and pilot whales make a variety of sounds that change in type and rate with behavioral activities.[21] For these species, the highest number and variety of sounds, including whistles, screams, and barks, are created during social activities. In contrast, when resting, dolphins can be nearly silent. Toni has found that captive bottlenose dolphins whistle more when swimmers are in the water than when they are not.[22] These observations seem intuitively correct—specifically that signal exchange is higher and more complex when dolphins are more interactive, just as a room full of socializing teenagers sometimes threatens to break the decibel barrier with whooping and hollering.

Although data on rates and occurrences of vocalizations are valuable, we must identify the individuals making specific sounds as well as examine the behavior that accompanies the sounds to develop a complete picture of dolphin communication. When observing how dolphins behave at close range, identifying which individual initiates contact behavior (touching) and which receives it is relatively straightforward. The same can be said of postures or other visual cues. As we have seen, however, dolphins do not have to be within sight of one another to communicate. This, coupled with the facts that sound travels four and a half times faster underwater than in air, that dolphins usually exhibit no external sign that they are producing sounds, and that humans are designed to localize sounds in air but not water, makes it hard for humans to assign initiator and receiver roles in dolphins in the wild. Belugas are unusual in this regard because we can often see who is vocalizing

The mobile video/acoustic system (MVA) is the tool that Kathleen uses to record and document dolphin signal exchange. The MVA captures stereo audio with a video record of the behaviors.

—the undulations of the melon are almost as visual as a person's moving lips. Belugas are also the only odontocetes that can move their neck to orient their head in different directions, which might aid in determining who is vocalizing. Other odontocetes cannot move their necks freely.

The advent of new technologies and a melding of talents among biologists and engineers have paved the way for novel methods for recording dolphin behavior and sounds simultaneously underwater. In 1992, with advice from my dad, Pete (an electrical engineer), and graduate adviser, Bernd Würsig, I (Kathleen) designed and built a mobile video/acoustic system (MVA) to record dolphin behavior and their sounds concurrently, which I have been using ever since as a tool to study dolphin communication. The MVA, or "array," as I affectionately call it, has two underwater microphones (hydrophones) spaced at roughly four and a half times the distance between my ears. The hydrophones are located at the ends of a bar attached to the housing, where they are plugged into a stereo video camera.[23] In 1997, after several years of underwater observations and much already learned about dolphin behavior and interaction, I added a third hydrophone, a digital audio recorder, and a circuit board to capture and record echolocation from wild dolphins as they approached the MVA.[24] This unit was nicknamed the ECD, or echolocation click detector, because it allowed us to record click trains from individual free-ranging dolphins that we were observing as they swam toward the camera. These two devices were, and still are, innovative even with only two or three hydrophones. It was not until

2005 that modifications to this original design were proposed and made possible because of more recent technological advances, such as the creation of really tiny hard drives and even smaller cameras and batteries.[25]

My array is one of the few tools available to us to study free-ranging dolphin communication from underwater. Individuals in a group may consistently surface to breathe in a reliable pattern, but when socializing underwater they zigzag around one another in what appears to human eyes as aquatic "tag." Each season and every new recording and observation provide fresh insights into what and how dolphins are signaling to one another. Observing dolphins that are habituated to swimmers in relatively clear water facilitates the study of dolphin communication: more often than not, we can identify the signals used, the sender, and the receiver by using the MVA.

Another tool of our trade is in our methodology for recognizing and identifying individual dolphins. We use naturally occurring scars and marks to recognize each dolphin or beluga reliably in each group that we study. Most dolphins are easily identified by the notches and scars in their dorsal fins. When studying them underwater, we gain access to views of the entire dolphin body and can document marks, pigment differences, and scars anywhere on their form. We assign each dolphin a name and/or number. Standard convention in the scientific study of animals dictates that we maintain objective and impersonal views of our subjects to prevent us from anthropomorphizing them. In contrast to this system, Jane Goodall named the chimpanzees at Gombe, Tanzania, as an acknowledgment that the chimps were individuals with character traits and "personalities"—this helped people (scientists included) to traverse the chimp-human border. Since Goodall's studies, the assignment of names to nonhuman subjects has become more accepted.

Several attempts have been made to inventory the whistle repertoire of wild and captive dolphins. The size of a repertoire, for a species or an individual, including both whistles and pulsed sounds, is probably limited to fewer than forty discrete types.[26] It is possible that whistles are graded, rather than discrete, signals—that is, these signals sort of fade into one another, having subtle differences in meaning rather than being distinct sounds. Although this is not a vocabulary like we would find in a dictionary, we are beginning to understand how some dolphin sounds might correlate to certain behaviors. We are learning about the "how" and the "what" behind dolphin communication. Consider again the importance of touch for communication. For instance, placement of the pectoral fin by one dolphin to the side between the dorsal fin and the flukes of a second dolphin is termed *contact position*.[27] Researchers in Australia have described this behavior as "bonding."[28] In most reported cases, the action is a request made by the first dolphin. In another example, pairs and triplets of dolphins engage in ganglike fights, "squaring off" with loud sounds, bubble emissions, and lots of posturing. After about five to six minutes, the fighting stops, and members of each pair or trio swim quietly in parallel in layered contact position.[29] While exchanging pectoral fin touches, the dolphins are nearly silent or whistling mildly. During the fight episode, between flipper touches and affiliative exchanges, the dolphins are noisy and overt in their actions. Therefore, "contact position" likely sends two messages—depending on the receiver. To partners: "We're still on the same team, right?" To opponents: "You still have to take on our team as a whole." These tactile messages are reinforced or solidified by the accompanying acoustics, postures, and gestures.

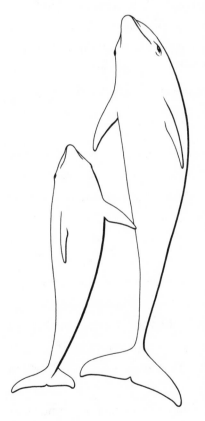

A clearer concept of the message behind the behavior of contact position as a request was

Dolphins place a pectoral fin on the side of another dolphin when asking for assistance or other action. We see this often between dolphins when they fight in groups. During breaks in fighting, dolphins within each gang place their flippers to the sides of their peers. This signal lets their partners know they are still a team and advertises to their opponents that their team is still tight.

strongly emphasized during my (Kathleen's) fieldwork in July 1994. It was dusk, with light too low for good video recordings and the sea just shy of turbulent (think of a washing machine on the gentle cycle). I was in the water with another woman and two juvenile female dolphins, Topnotch (ID#3) and Doubledot (ID#39). A glance at the boat, which was a good hundred yards (about 100 m) away against the current, told me it was time to say good night to our friends and battle the waves back to the boat. I signaled to my friend. We watched the dolphins swim out of view and then started our swim to the boat. After about a minute, both Topnotch and Doubledot swam back to us: Topnotch paralleled me, Doubledot my friend. As we swam, Topnotch let me place the back of my hand against her right flank—contact position. After we reached the boat, both dolphins swam back toward other swimmers finishing their last swim of the day. Topnotch then swam back to me, pressed her entire right side into my body, rubbed her belly into my belly, and then moved off to Doubledot! Had I asked her for the rub and contact? Or had I asked for some other signal? In that instant, more than any other in my life, I wished I had Dr. Doolittle's ability to talk to the animals. Maybe, with more observations of behavior, we will eventually be able to understand and converse, in some form, with the animals. ♡

Occasionally the observer becomes the participant, particularly when blinds (used to conceal the observer from land animals) are not an option for underwater dolphin research. When the animals being observed detect the researcher, challenges (or opportunities) arise. If the animals do not flee, the researcher, instead of being ignored, may become the target of attention. The observer may become an active participant in interspecies communication or may choose to be an unresponsive observer. Because dolphins are often so gregarious toward people, it is not uncommon for this to happen.

I (Toni) observed a free-ranging "friendly" beluga whale in Nova Scotia that had been interacting with swimmers and boaters in the area for several years. It was my first time in the water with this whale, and my intention was to make myself as uninteresting as possible to her so that I could study her underwater behavior as an objective observer. The whale, however, had a different agenda. It seemed that

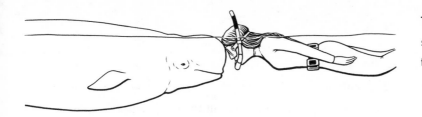

Toni and Wilma share a quiet moment together.

the less responsive I was to her solicitations for attention, the more persistent she became in engaging me to interact with her. After repeatedly rubbing her body along my hands, which I kept relaxed at my sides despite the temptation to stroke her, she reoriented her body at the surface so that we were both floating, head to head. She pressed her bulbous melon into my forehead ever so gently and just held it there, almost motionless.

Although I did not understand the specific meaning of this behavior, it seemed clear that she wanted me to interact with her rather than simply observe her . . . so much that she took matters into her own hands. I saw this behavior again with a different "friendly" beluga when he gently pressed and held his melon against a diver's head. I have been unable to find anyone who has observed this behavior occurring between belugas. Perhaps the rare opportunities for close, in-water interaction with these animals reveal aspects of their communication that we would be unlikely to witness by observing them from afar. In this way, and when conducted responsibly, close observation may provide an invaluable window into the rich communicative expressions of dolphins.

Individuals within species direct behaviors to one another that are generally understood and interpreted correctly by other members of their species (or culture, when applicable)—and the responses they receive are somewhat predictable, too.[30] As a result, members of a species typically know what they can expect from one another. Effective communication can certainly be more challenging between individuals of different species, however. Signals that have become specialized through evolutionary processes to communicate with peers are unlikely to be highly effective for communication across species.[31] And yet, because members of different taxonomic groups are often in contact, communication does occur between individuals of different species.[32]

Over time, individuals, whether dolphins or other animals, may learn the meanings of signals exhibited by members of another species, and the two species may even develop mutually understood signs.[33] Examples of this are replete in human interactions with domestic animals and in animal training. Naturalists and researchers also attempt to learn and exhibit the signals of other species in efforts to interact with them. This strategy has been used with dolphins by the pioneering cetologist Kenneth Norris and others, with chimpanzees by Jane Goodall, and with captive and free-ranging gorillas by Brenda Patterson and Diane Fossey, respectively, among others.[34]

Cetaceans of different species interact with one another more often than do many other taxonomic groups. Spotted dolphins in The Bahamas frequently intermingle with bottlenose dolphins.[35] Common and dusky dolphins travel together off Argentina as well as in New Zealand waters.[36] Mixed species aggregations are also regularly seen in the eastern tropical Pacific.

Dolphins that frequently interact with humans may have learned how to communicate on some level with people, especially individuals they interact with repeatedly. One example of this communication may be the dolphins' mimicry of human postures and vocalizations, as exhibited by free-ranging Atlantic spotted dolphins, lone sociable bottlenose dolphins, solitary free-ranging beluga whales, and captive dolphins.[37] Dolphins who are around people in swim programs often circle or somersault around and in copy of human actions. Interspecies communication may also occur on a different level between captive dolphins and their trainers during reinforcement training.[38] Dolphins learn the meaning of their trainers' signals and behave according to the information inherent in the signal. Conversely, trainers often learn the significance of "*intra*-specific social signals" that dolphins may direct toward the trainer.[39]

Communication may also occur through play behavior, which can serve as an important mechanism in the development of communicative skills.[40] For this reason, play between dolphins and humans may provide the motivation and basis for the development of mutually understood signals necessary for interspecies communication. Although play between species is not common, this may be due primarily to the relative infrequency of prolonged social interaction across species.[41] Of those species that do interact socially with others (think of domestic dogs and humans), play is common. Researchers suggest that play is a component of social interactions between free-ranging dolphins and swimmers who frequently engage with one another.[42] Dolphin behaviors such as allowing humans to ride or be towed and playing keep-away with seaweed and other objects are often documented.

So why study communicative behavior in dolphins? Obviously, it's a fascinating subject with many implications. But is there a greater purpose? Absolutely. Increased public education about dolphin communication and behavior may contribute to greater public protection of dolphins and their habitats. More directly, dolphin communication and expression convey important information about their internal states and subjective experiences. Knowledge of their behavior allows us to glimpse into their psychological and physiological condition.[43] And this information in turn enables people (including scientists, fishers, boaters, swimmers, veterinarians, and natural resource managers) to better manage and care for individuals as well as populations. For instance, what we have learned about dolphin behaviors associated with stress has been instrumental in setting policies for the tuna fishing industry, which has an incidental impact on dolphins. ⌊, not incidental

By eavesdropping on dolphins, we find that the ways in which dolphins communicate with each other—and even with us—are incredibly sophisticated and complex, even by human standards. Even with vast differences between humans and dolphins in physiology, sensory systems, and habitat, the language of dolphins is unfolding before us as we spend more time beneath the surface.

Man had always assumed that he was more intelligent than dolphins because he had achieved so much—the wheel, New York, wars and so on—whilst all the dolphins had ever done was muck about in the water having a good time. But conversely, the dolphins had always believed that they were far more intelligent than man—for precisely the same reasons.

DOUGLAS ADAMS, *The Hitchhiker's Guide to the Galaxy*

I have often seen dolphins playing keep-away with some oceanic toy. I've never joined in these games, remaining the objective observer. My first visit to the Kewalo Marine Laboratory in Honolulu, Hawaii, to film a segment for Dolphins *finally gave me the chance to play with a dolphin. Researcher Louis Herman brought out a ball and told me to toss it with Phoenix: I tossed it to her, and she slapped it back at me with her rostrum. We tossed back and forth a few times until she "faked me out" with a trick shot! I believe Phoenix had a sense of humor and planned to pitch a toss that I couldn't return. After this experience and many others like it, I have no doubt that dolphins, indeed many other social beings, have rich emotional lives that we occasionally are lucky enough to glimpse.—Kathleen*

TDI
dolphin
institute

We identify with other animals when their emotions, thoughts, and desires are clear and decipherable. As mammals, we share many behaviors with other mammals—especially large carnivores like ourselves. Their facial expressions, body postures, and vocalizations, though diverse, are similar enough to ours that we may correctly guess their state of mind. Our closest animal friends are most certainly dogs, highly social "pack" mammals whose behaviors we have come to understand after fifteen thousand years of cohabitation.[1] Reading a dog's behavior is second nature; few of us fail to comprehend the meaning of a wagging tail, a growl, or a playful doggy smile. In contrast, the behavior of a fiddler crab is wholly foreign to most of us; a claw-waving display, though meaningful to other crabs, makes little sense to people strolling on a beach. To understand the behavior of fiddler crabs requires careful observation and years of study—mammalian intuition does not get you far in the crustacean world.

Behavior, and in particular communicative behavior, provides humans (scientists and nonscientists alike) a window into the animal mind. A communication signal is, in most cases, an attempt by an animal to convey important information about its mental state to another animal. A dog's growl lets others know that the dog is angry or scared. In this way, a signal provides a direct line to the dog's mind and may allow us to predict its behavior.

There are dangers and pitfalls when it comes to interpreting the behavior of animals, especially when we are encountering animals whose communication signals

Sometimes it seems that dolphins have an amazing sense of humor. Could this be what this adult dolphin in Nassau thinks of the cameraman?

are less familiar to us. Dolphins display an array of behaviors that are unmistakably familiar, such as the joyful exuberance of a frolicking dolphin calf. But other behaviors are less easily interpreted and are often disastrously misinterpreted. Having swum with wild dolphins in the company of other people, we can both recall countless examples of misinterpretation by humans unfamiliar with dolphin behavior. An open jaw display might look like a dolphin laugh, but to the trained observer, this conveys agitation or aggression. In my (Toni's) research, I was surprised by the degree of interspecies *mis*communication that I observed. Decades of mistakes that I observed, from others and myself, often resulted in serious misunderstandings between dolphins and people. In one instance at Mikura Island, I (Kathleen) was in the water with another researcher and three female tourists. From the surface, we saw three to four dolphins socializing, but from underwater we realized that we were seeing two adult male dolphins corralling one subadult female. She was bleeding from marks left by their teeth. As the five of us watched the scene unfold, we inadvertently formed a semicircle in the water. The young female took advantage of this half-circle and used us as a shield against the angry males. She was not more than 3 feet (1 m) in front of my camera lens, and I could feel my heart pounding in my chest as I heard the loud pop sounds from the two impatient males. After about a minute, the female took off with the males in hot pursuit. As we got out of the water, the three female tourists were squealing with delight while the other researcher and I breathed sighs of relief. Even after we explained how dangerous the situation we just witnessed had been, they truly did not understand.

Unlike the smile of a dog, the dolphin smile is not a direct line to a dolphin's emotional state: it is a by-product of the structure of their jaw, a result of a morphological adaptation for sound reception. We need to look beyond the dolphin smile if we are to learn more about what is really going on inside their minds.

Relying on intuitive observation to decode the behavior of other animals will get us only so far; careful scientific study is required. For centuries the study of thoughts and emotions remained outside the reach of scientific method. In the early 1900s, the famed behavioral scientist B. F. Skinner left his mark on ethology by asserting that all animal behavior, including that of humans, is a reflexive, involuntary response to

The dolphin's smile is permanent . . . no matter what behavior they are engaged in, the smile will always be there.

stimuli—nothing more, nothing less.[2] Skinner's approach, however, ignores what most of us would consider an obvious point; behavior is governed by a complex mixture of thoughts, desires, and feelings. Charles Darwin, in *The Expression of the Emotions in Man and Animals*, placed great importance on the vital role that the communication of emotional states played in the evolution of species.[3]

The problem has always been how exactly to study emotions in nonhuman animals in a rigorous, scientific fashion. As animal researchers David Lusseau and M. E. J. Newman have pointed out, animals, unlike human subjects, "do not give interviews or fill out questionnaires."[4] Animal behaviorists do not have the luxury of being able to ask an animal what it is thinking. For many scientists this meant giving up on the idea that animals are even capable of thought or emotion. Yet today new methods and research tools are becoming available to help scientists identify answers to the questions surrounding the animal mind. Our toolboxes are expanding at an astounding rate, allowing researchers into a new and exciting realm of study and discovery. Dolphins in particular have provided us insight into the complex inner workings of the animal mind; dolphins have been involved in countless scientific studies that have produced remarkable results concerning their capacity for language comprehension, problem solving, and possibly even rational thought.[5] Observations of wild dolphins reveal highly complex social structures, tool use, and culture.[6] Thanks to a renewed interest in the study of the mind, we are finally starting to learn what is behind the dolphin's smile.

Dolphins have a popular reputation for being highly intelligent creatures. Modern movies and books often portray them as possessing incredible brainpower or an intelligence level just behind, on par with, or even ahead of humans.[7] Yet intelligence may be one of the most undefinable and nebulous subjects of study in modern psychology—so much so that the modern field of cognitive science (the study of the mental operations and tasks that are carried out by the brain) has little interest in placing animals on an arbitrary scale of intelligence. Instead, cognitive scientists are interested in determining what it is that brains are designed to do and then figuring out just how they do it. Modern scientific study of the animal mind seeks to describe the mental mechanisms that lead to observable behavior. These mechanisms may consist of emotions, desires, feelings, and thoughts, working in conjunction with memory, learning, and other machinery that are tasked with obtaining, processing, evaluating, and acting on information. The cognition versus intelligence dichotomy is really the difference between asking, "How does a car work?" and "Is the car fast?" The first question will provide you with concrete answers, whereas the second leads only to more questions, such as "Faster than what?" and "What does it mean to be fast?" In this analogy, the concept "intelligent" is akin to the concept "fast": both terms seem to defy concrete definition that would make them useful to scientific inquiry. It is more interesting to ask what it is that a dolphin brain can do and how it does it, than trying to determine whether dolphins are intelligent. Thankfully, the study of intelligence is no longer fashionable. Instead, the study of cognition and the mind is just getting warmed up.

What better place to begin the study of the mind than with the organ that gives rise to it—the brain? The dolphin brain has been an exciting topic for scientists because dolphin brains are big, bigger than the brains of most other mammals, and certainly those animals similar in body size to the dolphin. This fact alone created quite a stir in the scientific world some fifty years ago when facts about the dolphin brain started to surface. Dolphin research pioneers like John Lilly surmised that the physical size of the dolphin brain could be an indication that they are capable of impressive, if not fantastic, mental feats. These early days of dolphin brain research soon exceeded the capacity of the scientific methods of the day. Research in the past few decades has begun to paint a more complete picture of the dolphin brain and its structure.

One method of measuring brain size that goes beyond absolute size is the encephalization quotient (EQ). The EQ measurement is calculated by comparing the weight of an animal's body to the weight of its brain. Humans, not surprisingly, have a relatively enormous EQ, 7:1; this means that our brains are seven times larger than average for an animal of similar body size.[8] Many of our closest primate relatives, like chimpanzees, also have a large EQ, ranging from 1.5:1 to 3:1. In contrast, several species of odontocetes have an EQ ratio approaching 5:1, just shy of humans and well beyond that of the great apes.

Research conducted in the past decade sheds light on the course of dolphin brain evolution. The early aquatic ancestors of dolphins did not have particularly large brains. Around forty-seven million years ago, however, during a period known as the Eocene-Oligocene boundary, the Archaeoceti underwent a significant change: their brains grew larger as their bodies got smaller.[9] A second major change in brain size occurred fifteen million years ago in the Miocene era, when the odontocete superfamily Delphinoidea began evolving much larger brains than their other cetacean cousins. At the end of this period of drastic increase in brain size, odontocetes had a large brain that would hold the title of the largest EQ in history—before, that is, the arrival of *Homo sapiens.*

Simply having a large relative brain size is not necessarily an indication that complex processes are occurring within the brain. One controversial hypothesis suggests that the large brains of dolphins and whales evolved to function as internal radiators to keep them warm in cold waters.[10] Most scientists are convinced, however, that the structure of the dolphin brain itself, along with the observation of complex behavior, reveals that dolphins are cognitively complex animals.[11] Progress in the study of brain architecture, such as that conducted by researcher Lori Marino, brings new insight into the function of the dolphin brain. Observing differences in the dolphin brain's structure when compared to other mammalian brains is a powerful tool in understanding how the dolphin brain works.

Not surprisingly, the olfactory bulb—the part of the brain responsible for processing smell—is greatly reduced in dolphins, as are the areas in the limbic region of the brain related to processing information from smell (for example, the hippo-campus).[12] The corpus callosum, a series of connective tissues linking each of the brain's two hemispheres, is much smaller in dolphins than in humans. Dolphins, like humans, have brains divided into two halves, or hemispheres. Dolphin sleep physi-ology evolved in such a way to allow them to sleep one half of their brain at a time. The half that is awake controls breathing, swimming, and other behaviors while the sleeping half gets rest. Reduced connections in the corpus callosum make this pos-sible. In an aquatic environment, a dolphin that fell completely asleep might drown; unlike humans, dolphins are in conscious control of each breath they take.

Other structures in the dolphin brain are larger than that of other mammals. The auditory nerve is significantly thicker in dolphins than in humans.[13] This makes sense when we consider the enormous amount of information that a dolphin must transmit from the inner ear to the brain for the processing of echolocation. The area at the base of the brain that helps coordinate movement is also much larger in dolphins than other mammals, likely related to the need to govern dizzying body movements in a "zero gravity" aquatic environment.[14]

By far the most striking characteristic of the structure of the dolphin brain is the organization of the neocortex.[15] The neocortex is the outermost portion of the brain—the very top layer that rests on each of the two hemispheres. It consists of both gray matter (the cells responsible for information processing) and white mat-ter (the cells responsible for the transmission of information). It is the part of the brain generally understood to be responsible for higher cognitive functions includ-ing sensory perception, issuing motor commands, reasoning, thinking, planning, and so on. The human neocortex is highly convoluted, or folded on itself, creating the well-known wrinkled brain look. The wrinkling is the result of the need to fit a large neocortex into a limited skull space—a lot like crumpling up a large piece of paper to fit it into a small glass jar. By contrast, rats, which have a paltry EQ of less than 1:1, have a wrinkle-free neocortex. Dolphins actually have more wrinkles in their neocortex than do humans. The dolphin neocortex is not as thick as the human neocortex but is nonetheless impressively large.

The neocortex receives sensory information (for example, from sight, sound, or touch) in different areas known as projection zones. In general, projection zones follow a standard topography in mammal brains (such as vision at the back, touch at the top). Yet in dolphins the projection zones are spread out in strange areas of the neocortex and include a lot of associative cortex whose function is unknown.[16] Some scientists suggest that dolphins need this extra neocortical projection space for processing auditory information resulting from echolocation, although others strongly reject this idea.[17] The truth is that we are not really sure how the neocortex works for dolphins. Furthermore, the structure of the cortical cells themselves, their relative uniformity, and the way that they are arranged in the various layers of the neocortext can only be described as weird. There is still hot debate among researchers as to what this all means in terms of cognitive function (for example, does having poorly differentiated neuronal structure make an animal smart or stupid?).[18] In any event, it is clear that the dolphin brain has evolved to be both large and complex when compared to many other mammalian brains. Why, then, have animals like dolphins, humans, and chimpanzees evolved large complex brains but animals like rabbits and cats have not? We wonder if there must be something similar about the evolutionary pressures on these specific animals that funneled them down a path to large brain evolution. To understand more about what these pressures might be, and what a big brain can do, we need to explore the history of research on dolphin behavior and cognition. After all, it is the behavior of dolphins that has captivated the attention of scientists for decades.

Trained as a medical doctor and the author of many popular books, John Lilly is probably the best known of the early researchers interested in the study of the dolphin mind.[19] Lilly's experiments focused on the possibility of dolphin-human communication, operating on the principle that dolphins had what he termed "complex, highly abstract communication."[20] His writings and ideas are in large part responsible for our modern ideas of dolphins as superintelligent animals. Many of Lilly's methods, especially those employed in the 1960s and 1970s, were considered unconventional at best.[21] Still, the impact of his ideas on generations of budding scientists cannot be discounted, nor can the work of early pioneers like Kenneth Norris, who is largely responsible for the discovery of dolphin echolocation, or Karen Pryor, whose work with dolphin training and dolphin behavior began to suggest that there was something unique about the dolphin mind.[22] It was not until the discipline of cognitive science emerged that the study of dolphin behavior entered the modern era of scientific inquiry.[23] The cognitive revolution brought together disciplines such as psychology, philosophy, neuroscience, and computer science to provide the necessary framework from which the modern science of the mind could take shape.

Dolphins quickly found themselves the focus of experiments investigating their ability to both produce and comprehend symbolic communication. Like many great apes species, dolphins were tested for language comprehension in what would become a new line of scientific inquiry known as animal language research.[24] Animal language research remains a hot topic of research. Although language is an evolutionary adaptation often considered unique to humans, other species may be capable of a range of language comprehension and production.

The term *language* can be tricky to define: it is more than simply a complicated communication system. Biologists and botanists alike would point out that complicated communication is part of the life cycle of all animals and plants. Language is something else—an ability that is clearly tied to the capacity for complex thought. Many modern cognitive psychologists consider the ability to acquire and use human language as the result of an evolved, linguistic instinct that goes beyond a more general ability to simply learn a skill.[25] It involves cognitive processes that predispose human infants to learn how to combine meaningful sound elements using extremely intricate grammar. Learning what animals can and cannot do in terms of human language and symbol manipulation will tell us not only more about the cognitive processes of animals but about the likely course of human evolution and how and why we became the talking ape. Animal language research has at times been controversial, with scientists passionately debating the meaning and significance of their results.

Lilly began the first experiments to teach human language to dolphins. Although he was inspired by dolphins' vocal ability to mimic human sounds, these experiments were unsuccessful.[26] Other contemporaries, including Dwight W. Batteau, conducted language-training experiments with bottlenose dolphins, but without success.[27] Batteau, using Skinnerian stimulus-response shaping techniques and a device that transformed human words into underwater whistles, taught his dolphins to follow commands but could never claim that the dolphins were learning something akin to human language or that they were capable of symbol manipulation. About the time Lilly and Batteau's work was ending in failure, a breakthrough was made with another large-brained species, the chimpanzee. Two famous programs involved teaching sign-gestures to young chimps raised by humans. Allan and Beatrice Gardner, working with Washoe, and David Premack, working with Sarah, shocked the world when each announced that they had taught these chimpanzees to communicate with symbols.[28] Since then, many studies involving bonobos, gorillas, and parrots have reported success in teaching animals to use symbols to communicate.[29] Dolphins, too, are acknowledged as more than competent in this domain, thanks to the successful work of what are now at least two generations of researchers studying marine mammal cognition.[30]

Herman is perhaps the best known modern researcher of dolphin cognition. He and his team investigated the behavioral abilities of dolphins at the Kewalo Marine Laboratory in Honolulu. Herman's research differed from his predecessors' in teaching dolphins to comprehend symbolic communication but not to produce it. The stars of his research program were two bottlenose dolphins, Akeakamai (Ake) and Phoenix. Ake was trained to understand gestural symbols (given with the arms and hands), and Phoenix was taught whistle symbols. His achievements refocused the dolphin at the center of the animal language research debate.

Using gestural and whistle symbols, Ake and Phoenix were taught a system similar to language that involved both syntactic and semantic rules. In other words, the dolphins understood the meaning of the symbols, as well as how things like word-symbol order can affect meaning. For the dolphin to understand the ultimate meaning of a sequence of symbols, or sentence, the dolphin would need to wait until all the symbols were produced; only then would the meaning of the sentence be clear. The symbolic sentence *Surfboard → Swimmer → Fetch* is understood by the dolphin as "Bring the swimmer to the surfboard," whereas *Swimmer → Surfboard → Fetch* means "Bring the surfboard to the swimmer." Ake and Phoenix understood these sentences and carried out the appropriate actions, suggesting that they understood not only the meaning of each word-symbol but the importance of word order. Not all sentences were formed in this linear fashion; for example, the phrase consisting of the symbols *Basket → Ball → In* means, "Take the ball and put it in the basket." Ake and Phoenix had no trouble understanding this command.[31] But the phrase *Ball → In → Basket* means the exact same thing, though the words are in a different order. The dolphins understood this sentence, too. In 1984, when the results showing Ake and Phoenix's abilities were first published, they created a stir in both the scientific and nonscientific communities.[32] There was much controversy, but the work of Herman and his team is heralded today as a breakthrough for animal language research.[33]

This same team also reported discoveries that shed light on the mental machinery that may drive a dolphin's ability to manipulate symbols, including the ability to understand novel sentences and the ability to improvise solutions to sentences that are convoluted or nonsensical.[34] When Ake was asked to "bring Phoenix to the water," Ake simply stared back at the trainer, deciding to ignore what was apparently a stupid request. When given a sentence that violated a syntactic rule, like *Speaker → Water → Pipe → On* (in which three objects are confusingly named in a row), Ake tried to make sense of it. In this case, she fetched the pipe and placed it on the speaker, having decided to ignore the water symbol even though it was in the middle of the sentence. For her, it made more sense to bring the pipe to the speaker. When given a command to carry out actions on objects that were not in her pool, Ake would touch a paddle in her pool (the "No" paddle), indicating that she could not complete this request. If one of the objects was present (for example, the Frisbee) but another was missing, she would fetch the Frisbee and bring it to the "No" paddle. When presented with the command *Surfboard → Over,* Ake jumped over the surfboard floating in her pool. But when issued this command with the surfboard resting against the side of the pool, she would first bring the surfboard into the middle of the pool and then jump over it. Such responses indicate that Ake understood the concept of "over" and had enough foresight and problem-solving ability to rearrange objects in her environment to complete a command. Although these results do not suggest that dolphins will ever learn a true human language, this work with symbol manipulation by dolphins is considered some of the most successful research conducted on any animal species in terms of language-symbol comprehension. Such language experiments are, however, just the start of what we have learned about dolphin cognition.

There is more to cognition than manipulating symbols. Daunting numbers of publications on dolphin cognition detail everything from the function of the visual and echolocation systems to episodic and event memory to dolphins' ability to perceive and classify different kinds of objects.[35] We mention some of the highlights here.

Most animals have a difficult time interpreting what they see on television as analogous to what they perceive in the real world. Even chimpanzees do not quite understand what is happening when they are first exposed to television images. After training, they learn to perceive television images in a way similar to human beings. In contrast, dolphins seem to "get" television from their first exposure. In one experiment, a television was placed in the underwater viewing window of the dolphin tank, and a researcher gave gestural commands via the television screen.[36] The dolphin understood the commands right away: this ability is interesting if you consider that the image on the television was a tiny two-dimensional black-and-white representation of a human being! That dolphins could understand this image as representative of real-world situations reveals that they have a complex visual recognition system and an ability to interpret distorted visual stimuli.

Karen Pryor and Louis Herman have documented the dolphin's ability to create novel behaviors on command.[37] During training episodes, bottlenose dolphins (for Herman) and rough-toothed dolphins (for Pryor) learned that they would be rewarded for displaying behaviors their trainers had never before seen. Herman's team also devised a signal that asked a pair of dolphins to carry out behaviors in synchrony, something dolphins have no trouble with (synchronous behavior in the wild is quite common).[38] When this command was given in conjunction with the command to create a novel behavior, both dolphins "invented" a novel leaping behavior and executed it in perfect synchrony.[39] How they accomplished this is a mystery. The ability to execute synchronous behavior, coupled with the dolphin's known propensity for imitation, may in part explain why dolphins are so adept at producing complicated behavior (for example, mud-bank fishing or sponge feeding) that requires either close cooperation or the ability to learn from one's peers.[40] While working on a segment of *Dolphins* at the Kewalo Marine Laboratory, the movie's producers wanted to film tandem dolphin leaps. Herman requested a novel, synchronous leap as everyone watched. Kathleen was watching the dolphins from above the pool and saw them "converse" a split second before they leaped and spun to the left. That is, their rostrums were pointed at each other as they swam slowly next to each other just before they leaped. Were they vocalizing to each other? Did one suggest the type of leap and the other say "okay"? We did not have a hydrophone in the water, so we can only guess at the answer. The behavior suggests, however, that they communicated something.

In another display of cognitive prowess, dolphins have been found to comprehend human pointing gestures.[41] This is significant given that most scientists previously assumed that understanding a "point" was an ability unique to primates, especially given that dolphins lack arms and hands. The team at the Kewalo Marine Laboratory gave dolphins commands to perform an action on an object, but instead of naming the object using a symbol, the researchers simply pointed to the object. Dolphins can understand the reference of the human pointing gesture in a variety of forms and situations, and what's more, they do it without being expressly trained to understand pointing. Other researchers have obtained similar results in their experiments with dolphins.[42] This ability is rare in the animal kingdom—not even chimpanzees can understand pointing gestures this easily, although dogs seem to have a somewhat easier time than chimps.[43] There is even some evidence that dolphins produce pointing gestures of their own, using their rostrum.[44] What's more, they appear to alter their pointing behavior based on whether a human paying attention is present; when it seems to the dolphins that nobody is looking, they don't point as often.[45]

I (Kathleen) recall a situation of pointing among Indo-Pacific bottlenose dolphins around Mikura Island.[46] During a survey, we came on a group of dolphins that were spending a lot of time below the surface. Once I entered the water to observe them, I saw what held their attention: the dead body of one of their peers resting on the sea floor. For more than two hours, we watched as more than nineteen dolphins (all

Kathleen records four Atlantic spotted dolphins with
her MVA system. We prefer that the dolphins ignore
us when we observe their behavior.

Dolphins love to play. They'll play with each other,
with people, and with objects like this seaweed.
Sometimes they play keep away with prized toys.

Sargassum and other seaweeds often make nice aquatic jewelry for the dolphins. Play doesn't have to be the only function for these watery delights.

⬆ Killer whales travel in family pods; the adult males are readily identifiable by their tall dorsal fins. The adult male dorsal fin is at least twice the size of the adult female dorsal fin.

⬇ Spotted dolphins in The Bahamas are often seen rubbing part of their body into the sand—maybe to scratch an itch, rub off sloughing skin, or just because it feels good.

The strongest bond among most dolphins is that
between mother and calf. Calves travel near mom's
dorsal fin or just ventral to her for protection and
assistance during swimming.

➡ The biggest threat to dolphins worldwide is pollution. This plastic bag might seem decorative, but it is dangerous to a variety of ocean animals.

⬇ In a remote part of Mexico, Toni is approached by a "friendly" gray whale. The unique whale-human relationship that has developed here is staunchly protected and provides perhaps the only situation where whales can safely approach people for close interaction.

Human beings share the planet with thousands of
other species. We must learn to interact responsibly
with other animals and to share our planet's
resources.

Swimming with dolphins is an amazing experience.
Every country has different rules that should be
checked out before you dive in.

← Spotted dolphins are born without spots and gain pigmentation as they age. Here a swimmer views four spotted dolphins of different ages.

↓ If we are not careful, we might love these animals too much. Solitary, sociable toothed whales, like Wilma, often take humans as their group.

It is rare to find dolphins alone. They are often touching or rubbing each other.

⬆ Dolphins already share a set of communicative signals with people: they understand both gestures and sounds from human trainers.

⬅ Dolphins are voluntary breathers and exhale just before their blowhole breaks the surface. The air bubble from this mom is a good example of a dolphin exhalation.

Dolphin watching has grown exponentially around the world.

← At times it is hard to determine just who is watching whom.
⬇ Mutual respect allows a trainer to request a specific behavior from a dolphin. Husbandry behaviors are important to the well-being of many animals in captivity.

⬆ Like most social animals, the young dolphins are most playful.

⬇ Belugas have a bulbous forehead that seems to express their inquisitive nature.

Dolphins have a reputation for being a "happy-go-lucky, friendly, helper to man," which their smile supports. Dolphins are social mammals and can be aggressive.

Caught! Sometimes the dolphins are quite inquisitive
of our observations . . . or maybe just of us. Here the
tables are turned on Kathleen as she watches two
bottlenose dolphins around Mikura Island, Japan.

but two of them male) swam repeatedly to the carcass and echolocated or pointed at the chest and the genitals. We learned later that the dolphin had drowned and hypothesized that the attending dolphins were checking out the lungs—their pointing gestures were overt and likely kept everyone (dolphins and observers alike) looking at the same area.

Pointing is an important subject for cognitive scientists because it suggests that dolphins may be aware of the minds of other animals. By pointing, I am actively trying to get *you* to focus *your* attention on something. Thus, I must realize that you have a mind that is capable of changing focus. Some researchers suggest that dolphins may have a predisposition for understanding attention because of a unique ability that results from echolocation.[47] In one experiment, a dolphin, positioned so that she could listen in on the echolocation activity of another dolphin but not produce echolocation of her own, was able to use the information she heard in the other dolphin's click echoes to perform an object choice task.[48] For example, if the first dolphin echolocated on a ceramic mug (without being able to see it because it was hidden behind an opaque barrier), the second dolphin, listening nearby, picked out the mug when asked "Which of these two objects did you hear?" Could dolphins in the wild in fact be eavesdropping on each other's echolocation, receiving all sorts of information about the environment without having to do anything more than listen? If dolphins are able to understand that echoes are a result of the echolocation of other dolphins, then they might also realize that the object of another dolphin's echolocation is the same object that they are hearing. This psychological concept is called joint attention, and sensitivity to joint attention might explain why dolphins are so adept at understanding pointing. Dolphins may be born with a keen awareness of other animals' states of attention.[49]

The disciplines of psychology, philosophy, biology, and neuroscience have grappled with the nature of consciousness and self-awareness for centuries, and they have yielded an incredibly diverse collection of opinions on the matter. Depending on whom you talk to, consciousness has either already been adequately explained or can never be adequately explained. It may be present only in humans or it is likely to be found in nonhuman animals as well.[50] Like most problems of this nature, the main stumbling block is to establish a working definition. A medical definition of *consciousness* would perhaps distinguish between a state of wakefulness and a state of sleep: in this sense, any animal that is awake is, by definition, conscious. But a philosophical-psychological definition of consciousness involves the extent to which one has awareness of one's own mind, and it can include any number of ill-defined variables, involving descriptions of what it means to experience things subjectively: the qualitative feeling of phenomena. These concepts are all but impossible to study empirically, especially in nonhuman animal species like dolphins that cannot tell us how (or if) they are feeling. Leaving aside the more difficult problems of consciousness, cognitive scientists have, despite initial difficulties, made great strides in tackling the issue of self-awareness. The problems of self-awareness, if understood as the question of whether an animal knows what it is thinking and whether it is aware of its own uniqueness in the world, are now understood to be questions to which science *can* provide concrete answers. For dolphins, many of these answers have been downright remarkable.

A basic test of self-awareness is to determine ability to understand the location of one's own body parts. Researchers at Kewalo Marine Laboratory assigned gestural symbols to different parts of the dolphin's body.[51] The trainers used these gestures to ask dolphins to do different things with their various body parts—for example, "Touch the Frisbee with your dorsal fin." The dolphins easily completed these tasks. In addition, dolphins imitated human behaviors by learning to associate their own body parts with those of the humans.[52] For example, if the human trainer bobbed her head, the dolphin would also bob her head. Forming an even more complex analogy, if the human waved her feet, the dolphin would wave her tail. Other than humans, no other animals that have been studied can imitate body movements in this way. Dolphins have long been considered highly skilled mimics: they are able to imitate both familiar and novel actions produced by humans or other dolphins.[53]

Even more impressively, research has documented that dolphins can recognize themselves in a mirror.[54] The mirror self-recognition test, pioneered by psychologist Gordon G. Gallup, Jr., with chimpanzees, is used as a direct test of an animal's capacity for self-awareness.[55] Gallup's experiment works basically like this: You give the test subject(s) a few days to familiarize themselves with a mirror. During this time—if the animal is up for the task—it will come to realize that the mirror is a reflection of its image. Most animals will display social behaviors at the mirror as if they were encountering another animal. But if the animal manages to catch onto the concept of the mirror, any social behavior will stop. After the animal is familiar with the mirror, researchers give the animal some kind of visible mark on its body. If the animal appears to inspect the mark on its body with the mirror, then voilà—you have mirror self-recognition. This experiment has been tried successfully with elephants and the great apes (humans, chimpanzees, gorillas, and orangutans).[56]

In 2001, Diana Reiss and Lori Marino performed the mirror self-recognition test on two dolphins at the New York Aquarium. After having been exposed to mirrors, the dolphins were marked with temporary ink. When marked, the dolphins twisted and turned and tried to look at their marks in the mirror, exhibiting behavior similar to chimpanzees when inspecting their marks. The dolphins thus understood that the animal in the mirror was in fact an image of themselves and not another animal. Earlier experiments have suggested that dolphins distinguish between watching themselves in mirrors and watching video recordings of themselves, leading some researchers to conclude that dolphins are self-aware.[57]

The behaviors discovered in the self-recognition experiments might be the result of something more important than an ability to figure out what a mirror does. Many psychologists think that recognizing yourself in a mirror is the first step in a chain of abilities that leads to a "theory of mind," a concept intimately related to the notion of self-awareness. Theory of mind is a psychological ability that works like this: I know that I have my own mind with my own thoughts, beliefs, and desires and consequently that you must have your own mind with your own thoughts, beliefs, and desires. Having a theory of mind is a powerful thing, because it allows an animal to predict and manipulate the behavior of other animals. It also leads to advanced social emotions like empathy. So if an animal like a chimpanzee, elephant, or dolphin can recognize itself in a mirror, this might mean that it is self-aware. If it's aware of itself, then it might be aware that it has its own thoughts, which means it might also be aware that other entities may also have their own thoughts: a theory of mind. Of course, it may be the case that the mirror self-recognition test is not the smoking gun that proves that animals have a theory of mind. It could simply mean that the animal subject figured out some basic properties of mirrors without making any giant cognitive leaps.

An even more advanced test of self-awareness and theory of mind comes in the form of the false belief task. The false belief has long been used to test children's ability to understand that other people can have beliefs that are different from their own and that these beliefs can be false. It works like this: an experimenter places a reward in one of two containers in full view of a child and a fellow observer. The observer then leaves the room, and the sneaky experimenter switches the reward from one box to the other. When the observer comes back into the room, the child is asked which box the observer should look in for the reward. If the child understands that the observer was not there to witness the switching, then he or she will know that the observer will hold the false belief that the reward is in the wrong box. When asked to point to the box that the observer thinks the reward is in, the child will point to the first box, even though he or she knows this is the wrong box. Human children are able to understand this concept beginning at age three or four, but not earlier.[58] Chimpanzees and other great apes generally have a difficult time with the false belief task. In one trial of the false belief test involving dolphins, however, the dolphins understood that the human researcher falsely believed that the wrong box contained the reward.[59]

The study of self-awareness in dolphins is just getting started. Experiments concerning the cognitive abilities of dolphins are revealing the extent to which the big dolphin brain is responsible for complex cognitive behavior. We are both excited about what the future holds; startling discoveries are likely to be made at every turn. Our own research has led us to study dolphin behavior outside a laboratory setting, to investigate how the big dolphin brain produces behavior in a natural setting. The discoveries concerning the nature of the dolphin mind as witnessed in the lab have many parallels with the complicated behavior we regularly encounter in the wild.

Dolphin performance in a laboratory setting has made it clear that they possess a complicated array of cognitive mechanisms. Dolphins can work with abstract concepts, plan ahead, monitor and regulate their behavior, imitate the behavior of others, and learn complicated concepts. All of these facts suggest that dolphins, perhaps more than other animal species, are prime candidates for having evolved another complicated behavioral process: culture. The term *culture* has a variety of definitions. The following description is perhaps most useful to our discussion: culture is information or behavior acquired from others through some form of social learning.[60] Social learning can include any number of concepts, but it basically involves the ability of one animal to acquire a new skill or behavior by observing, or being taught by, another animal. Social learning leads to behavior that cannot be explained by the normal methods of behavioral evolution, gene mutations or individual learning. Culture and social learning produce behaviors that are passed down from generation to generation or passed horizontally among individuals of the same generation. Unlike other kinds of behavior, actions arising from culture will be extinguished if those individuals who can pass it on are removed from the population. Social learning is most readily apparent in animals with complex societies and advanced cognitive abilities.[61] If that is the case, then dolphins are prime candidates for having culture.

Dolphin societies are complex, differ greatly from species to species, and vary among different local populations of the same species. As we have discussed in earlier chapters, many species of dolphin live in a fission-fusion society, regularly changing social partners. Of all the research conducted, the most detailed accounts of dolphin society are offered for two study populations: killer whales in the northeastern Pacific and bottlenose dolphins in Australia.[62] These two populations live in very different and somewhat juxtaposed social structures, and the killer whales have a particularly unusual social life.

Killer whales in the northeastern Pacific live in stable, matrilineal groups; young orcas spend their entire lives within their mother's pod (see chapter 2). In the population of resident orcas found in the waters off British Columbia, individuals within a matriline produce vocalizations, known as discrete calls, that are unique to that group.[63] These calls vary from group to group and pod to pod, which has given rise to the idea that orcas have vocal dialects.[64] Pods produce an average of eleven discrete calls; the calf learns these calls during the first few months of life. The structure of the shared calls changes over time, and a calf may learn to produce the calls in a slightly different way to that of its family members.[65] Over time the shared call structure of the pod changes, similar to the way human accents change over time. As each matriline follows its own path of vocal evolution, its calls will change over the years, eventually resulting in completely different dialects. Parallels with humans are obvious; it took only a few hundred years for the accent of the original English settlers in North America to evolve into the rich tapestry of American accents that can be heard today from Maine to Georgia to Texas.

Similar dialect traditions are observed in sperm whales. Sperm whales, like killer whales, live in long-term social units and produce stereotypical calls that are unique to that unit.[66] These calls, termed codas, consist of a series of click sounds and are most often heard during social encounters.[67] Researchers studying sperm whales have argued that, as with orcas, these coda dialects are transmitted through culture.[68]

The ability to learn vocal dialects requires a flexible system for vocal imitation. Killer whales clearly have the capacity to learn the discrete calls that will allow them to communicate with pod members. This flexibility also allows them to copy the calls of unrelated clans—mimicking the calls of other groups.[69] In one example, a captive orca originally from Iceland learned to produce the entire vocal repertoire of its tank mate (originally from British Columbia) after just a few years of living together.[70] In another study, a young captive killer whale calf learned its mother's calls even though it was housed in a tank with a third orca.[71] Although exposed to a flurry of vocal activity from both adults, the calf selectively learned its mother's calls while seemingly filtering out all other calls.

Cultural transmission of behavior in killer whales does not end with dialects. Killer whales near the Crozet Islands in the south Pacific, as well as those off Argentina, are regularly observed feeding on breeding colonies of sea lions that haul out on the beach.[72] To catch the sea lions, hunting orcas will intentionally strand themselves, rushing onto the shore to grab their prey under the cover of an incoming wave. It is a dangerous technique. To learn this dangerous technique, mothers spend years teaching their calves how to strand safely. Adults push the calves up onto the beach, directing them toward the sea lions, and help dig them out if they get stuck. Mothers and calves practice beaching themselves in the surf even if no sea lions are in sight. One study showed that calves who had a mother who spent more time teaching them these skills became much better hunters far earlier than their peers who received less training.[73] These kinds of active, long-term, and highly directed learning behaviors are rare in the animal kingdom.[74] Concerning the cultural traditions of killer whales, in terms of both dialects and other socially learned behaviors, Luke Rendell and Hal Whitehead state that "we are aware of no phenomena outside humans comparable to the distinctive, stable, and sympatric vocal and behavioural cultures that appear to exist at several levels of killer whale society."[75]

In Shark Bay, western Australia, Richard Connor, Janet Mann, and their colleagues have been studying a group of Indo-Pacific bottlenose dolphins, the same dolphin species that Kathleen studies around Mikura Island. The Shark Bay dolphins represent a local resident group of more than six hundred individuals and have been observed regularly since 1984.[76] Much like the Mikura population, these dolphins live in a classic fission-fusion society. Over the years, researchers have made a variety of startling discoveries that suggest an extremely complicated social structure. One discovery is the concept of the male alliance: two or three males who form a long-term alliance, cooperating to herd females during mating periods. Some male alliances are so close that the dolphins spend nearly every waking second in each other's company; alliances can continue for up to twenty years.[77] Male dolphins do not form such alliances haphazardly: they have very strong preferences about whom they will associate with. These small "first-order" alliances often combine with a second alliance group, creating what is termed a "second-order" alliance. Second-order alliances rely on one another to defend against other second-order alliances and to bolster their numbers when herding females. These alliances generally do not last as long as first-order alliances, usually no more than a few years.[78] Sometimes former members of a second-order alliance team will end up on opposite teams in future conflicts. Even more complicated alliances are also formed: a super-alliance of fourteen or more dolphins can greatly increase the courtship success of each individual in the alliance. These super-alliances are much more prone to changing membership.

Multilevel alliance relationships (first-order alliances, second-order alliances, and super-alliances) are an indication that dolphins have the mental machinery required to mediate such complexity.[79] The size of the neocortex might be correlated with the number of social partners that a species typically keeps: the more relationships to navigate, the larger the brain.[80] The mental muscle required to navigate such social complexity should not be underestimated. A complex hierarchy of this nature contains various "nested" alliances; individual dolphins must be able to keep track of where all the other individuals stand within each hierarchical level. Given the constant state of flux that some of these alliances are in, this is no mean feat. Humans routinely navigate complex nested alliances, but we are, after all, the animal with the highest EQ.[81]

Richard Connor and his team at Shark Bay have offered an explanation why dolphins evolved to be such large-brained creatures displaying such behavioral complexity. They suggest that the need to mediate such multifaceted, fission-fusion social networks requires an animal to evolve more powerful cognitive abilities.[82] This idea is termed the Machiavellian intelligence hypothesis—that is, the key to understanding why dolphins have complex brains may lie in understanding more about their social lives.[83] A similar theory has been offered to explain the evolution of a large brain and complex behavior in humans.[84] These ecological pressures may be even more pronounced in an ocean environment. Connor and Janet Mann point out that "cetaceans at sea inhabit a three-dimensional environment in which they cannot climb up a tree, crawl down a burrow, or hide behind a rock; they have nothing to hide behind except each other."[85]

The team at Shark Bay has produced even more remarkable discoveries. As early as 1997, researchers observed a handful of individual dolphins carrying sponges on their rostrum tips, apparently using them while foraging.[86] This "sponging" behavior was the first reported case of tool use by dolphins. The dolphins apparently use these sponges like gloves to protect the sensitive skin on their rostrums. As they poke around the ocean floor looking for food, the sponges protect them from objects, including animals with stingers like the stonefish. They sometimes also use the sponges to stir up the bottom, flushing out potential prey. Tool use is now realized to be much more widespread in the animal kingdom than was once thought: chimpanzees, monkeys, and crows, for example, have been documented as proficient tool users.[87] Given that dolphins lack hands and fingers or a particularly nimble beak, these observations of tool use in Shark Bay are remarkable. Unexpected as it may be given their morphology, the ability to manipulate tools seems well within the cognitive abilities of dolphins.

More surprising, however, is not how these dolphin spongers go about sponging but who is sponging: females almost exclusively. In an article about Shark Bay dolphin spongers in 2005, researchers reported that only one sponger was male, whereas thirteen were female.[88] The Shark Bay researchers suggest that the sponging technique is passed down from mother to daughter—that is, along a matriline. Comprehensive analysis ruled out a genetic component for the behavior. And because sponging and nonsponging dolphins forage for fish in the same environment, the behavior cannot simply be attributed to their ecology. Young dolphins likely learn the sponging skill from their mothers and then pass it along to their offspring. Thus, tool use among dolphins is another example of cultural transmission. Genetic analysis suggests that this skill arose recently and that there may have been a lone female dolphin who first came up with the idea of sponging—a "sponging Eve" of sorts.

Further evidence of something akin to cultural transmission in Shark Bay includes the phenomenon of hand feeding. At Monkey Mia beach, one group of dolphins regularly solicits food from humans. Not all the dolphins of Monkey Mia engage in this behavior, but the offspring of dolphins who hand feed also themselves solicit food from humans, suggesting that this is a learned behavior passed from mother to offspring.[89] Other dolphin populations engage in foraging behavior that appears to be passed down through generations. In Laguna, Brazil, a group of bottlenose dolphins has been working with local fishermen to herd fish into their nets since 1847.[90] The dolphins herd the fish toward the fishermen and execute a distinctive rolling dive that tells the fishermen when to toss their nets. Other groups of dolphins living in the area do not participate in this group hunt; only dolphins whose mothers were involved engage in this behavior, which suggests that only this one group has adopted this cultural tradition.

Cultural traditions in dolphin societies may in fact be commonplace. In attempting to explain the vast array of highly variable dolphin behavior that is witnessed all around the world, scientists are beginning to take the notion of culture in cetaceans much more seriously.[91] As the above examples from killer whales and bottlenose dolphins make evident, many behaviors are difficult to explain without invoking some level of cultural transmission. A requirement of culture is a brain capable of processing the kinds of behaviors that support social learning, especially given the demands of an intricate social network. Dolphins, much like primates, have evolved a brain that is similar to the primate brain in many respects, and it is no surprise that the social behavior of primates and dolphins have many parallels.[92]

In *The Descent of Man, and Selection in Relation to Sex* (1871), Charles Darwin wrote that "happiness is never better exhibited than by young animals, such as puppies, kittens, lambs, etc., when playing together, like our own children."[93] As we noted at the start of this chapter, humans tend to empathize most deeply with animal species that engage in behaviors that we recognize in ourselves. Like many of our fellow mammals, both humans and dolphins regularly engage in play behavior. Scientific reports of dolphins involved in complex play abound; their propensity for blowing and playing with bubble rings is widely cited.[94] Some researchers have argued that the complex play behaviors exhibited by dolphins require a level of coordination that is indicative of something akin to culture.[95] Using our built-in mammalian behavior detector, any human would conclude that a playful dolphin is a dolphin experiencing the emotion of joy. Is this really the case? What can play behavior tell us about what is happening in the dolphin mind?

The study of dolphin cognition suggests that the dolphin mind is capable of many complex activities: problem solving, planning, manipulating abstract symbols, and, in a more general sense, *thinking*. Scientists have moved beyond the Skinnerian approach to behavior and have concluded that the animal mind is indeed a mind full of thoughts, and perhaps at times even a rational mind.[96] But does the dolphin mind *feel* as well as *think*? Is their playful behavior an indication of their ability to experience emotions? As Darwin pointed out, the exuberant play of young animals appears at times identical to the playful behavior of our own children. This conclusion, however, is largely intuitive on our part—a gut feeling that takes you only so far. To describe the nature of emotion in animals, the scientific method is our best tool. Unlike cognition science, the scientific study of emotion in animals is somewhat controversial, often labeled "soft" science and, even worse, unscientific.[97] Yet the science of the animal mind has begun to confirm what our gut feeling told us all along: the *thinking* animal mind may indeed be a *feeling* animal mind after all.

Emotions, as mentioned earlier, are a vital part of the vast array of mechanisms involved in the brain's regulation of behavior. Primary emotions like fear keep an animal on its toes—wary of dangerous situations. All mammalian brains, as well as the brains of most complex vertebrates, have a limbic system: a complex arrange-ment of neuronal systems in the midbrain.[98] Primary emotions like fear appear to originate here, and the limbic system is also responsible for organizing motivation, emotion, and emotional memory.[99] Within the limbic system, the amygdala—two almond-shaped areas on either side of the brain—processes emotional reactions to events and regulates memories in relation to these events. The amygdala is vital for our ability to function; damage to the amygdala destroys many animals' ability to process pleasure, pain, and fear, take any interest in caring for offspring, or even participate in any form of social communication.[100] Like humans, dolphins have a large, well-developed amygdala.[101] This fact alone is enough for any scientist to conclude that dolphins, like almost all animals with a limbic system, possess a brain that generates fear and other primary emotions. Why then do some scientists still insist that animals may not experience fear?

The trouble starts with the notion of what it means to experience a feeling. The minds of other beings (human and nonhuman) are essentially private minds.[102] As a human, I have two options for knowing that another human is experiencing an emotion similar to the emotions I experience: (1) they can tell me about their emo-tion, and (2) I can see in their behavior that they are likely to be experiencing an emotion. Obviously, animals cannot tell us what they are feeling, so we are left to evaluate the usefulness of option 2 in figuring out what is going on in their minds. Recognizing that a human or an animal is afraid is straightforward enough: changes in muscle tone, eye size, whimpering, and sweating all indicate that the emotion of fear is coursing through the brain. But we still do not know whether the animal (or human) is subjectively experiencing fear. That is to say, we are not sure that they are aware that they are afraid.

This is where things get tricky. Without a definitive tool to test whether an ani-mal experiences an emotion, there is plenty of room for skepticism. Historically,

At times dolphins may leap just for fun . . . and sometimes leaps become part of a game of keep-away or tag. Spinner dolphins are named for their spinning leaps.

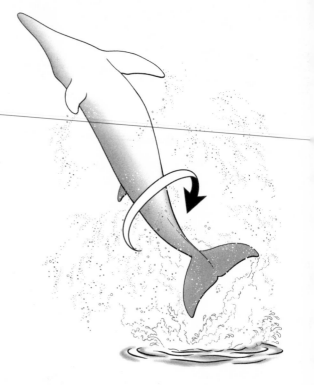

the norm has been to conclude that, because we cannot test for this subjective response, animals are not experiencing primary emotions, let alone complex secondary emotions like grief and embarrassment. To conclude that an animal experiences an emotion like a human simply because it appears to be experiencing such an emotion is anthropomorphism, the unjust attribution of human traits to animals.[103] Being guilty of anthropomorphism is something that animal behavior scientists are traditionally keen to avoid.

Not all scientists adopt this traditional stance regarding emotion in animals. Marc Bekoff argues that "those who claim that few if any animals have deep, rich, and complex emotional lives—that they cannot feel such emotions as joy, love, or grief—should share the burden of proof with those who argue otherwise."[104] In his view, we should ask, "Why must science assume that animals do not lead emotional lives in the first place?" If there is no evidence either way regarding emotion in animals, why prefer one stance over the other?

Thanks in part to the expanding toolbox of the cognitive revolution and a more scientific approach to the study of emotions in animals as advocated by Bekoff and others, scientists have made considerable progress in this area. Researchers have produced an enormous amount of data suggesting that there is good reason to accept that animals, like humans, have subjective experience of their emotional states.[105] As we have seen in the many studies of cognition in dolphins, they may very well be self-aware and, consequently, highly aware of their own emotional experiences.

Well-known reports concerning elephants, a large-brained social mammal whose matriarchal societies are not dissimilar in structure to many cetaceans (for example, killer and sperm whales), suggest that they display compassion and may assist other elephants in distress (both close kin and distant relatives).[106] What's more, they appear to have an awareness and curiosity of death and to suffer prolonged or permanent emotional damage reminiscent of post-traumatic stress disorder, as well as to produce abnormal or asocial behavior after exposure to severely traumatizing situations (such as witnessing the violent death of a family member).[107] As Gay Bradshaw and her colleagues note in their landmark study of this psychological disorder, young elephants are particularly sensitive to such trauma. The removal of a matriarch from an elephant herd may severely damage the group for several generations, because inexperienced female elephants cannot rely on help from an older, more knowledgeable mother; a similar process occurs when key individuals are removed from close-knit orca family groups.[108]

When I (Toni) first heard about this work, I was relieved because it explained to me some of the signs of stress and distress in dolphins that I had observed. Twenty years ago, it was academically adventurous to use phrases like "behavioral indicators of internal states in dolphins" to describe what I now call emotions. Just recently, I was hired to assess the "psychological state" of a young dolphin who had been moved from one facility to another, from one social group to another. Government officials could not understand why the dolphin was not socializing with her new pod. Emotional stress caused by an imposed shift in social group or anxiety caused by separation from her mother are ideas now gaining acceptance in mainstream science. I have observed similar emotional expressions in free-ranging solitary orcas and belugas prematurely separated from their mothers by death or geography. Trying to understand and assist these animals gives us the incentive to look deeply into the minds of dolphins, which may allow us a glimpse of what they experience in their hearts.

This field of inquiry is in its infancy. Most accounts of emotions in animals are by and large anecdotal, yet these accounts should not necessarily be dismissed as "unscientific." As Bekoff often points out, "The plural of anecdote is data."[109] Our own anecdotal accounts of dolphins behaving in what appear to be emotionally complex

Elephants share a surprising number of characteristics with cetaceans. In the company of experienced elephant researchers in Kenya, Toni observes their communication and social behavior up close without causing much disturbance.

ways are numerous—spending any amount of time with dolphins will inevitably result in such stories. Remember Kathleen's observation of the dead dolphin with several individuals pointing their rostrums at the chest and genitals of the carcass? The overwhelming initial response from several colleagues who viewed the video of this event was that the dolphins were attending a wake. The attending dolphins did seem to be grieving, but how do we define grief in a nonhuman animal when grief takes so many different forms among human cultures? The more we observe and document, the more refined our questions will be and the closer we will get to understanding the emotions of dolphins and other animals.

An unprecedented amount of research on the nature of the animal mind has been conducted in recent decades. Cognition, culture, and emotion—topics once thought outside the reach of science—are now fair game as our ability to study the mind increases. Dolphins are revealing themselves as highly social, complicated animals with a lot going on in their enormous dolphin brains. Perhaps some of our more fanciful ideas of the talking dolphin or the superintelligent dolphin are being reined in by a more responsible application of the scientific method. But the new reality of the dolphin mind is in some ways even more remarkable than what researchers dreamed of half a century ago. Scientists have begun cataloging an impressive array of abilities for dolphins—self-awareness, problem solving, abstract thinking, tool use, and social learning—abilities that are no longer considered unique to humans. Teasing apart the inner workings of the dolphin mind is enormously complicated, requiring as much thought, analysis, and philosophizing on dry land as time spent working with the dolphins in the water. For the problem of the dolphin mind, it is not a case of the devil being in the details but in fact quite the opposite. The details are where we have discovered the best bits of all. Having dedicated so many years of our lives to the study of dolphins, we could be ashamed to admit what is inescapably true: we still understand very little about dolphin societies and the dolphin mind. This is not reason for despair but cause for hope. We have many years of exciting effort as we begin to unravel the mysteries of the dolphin mind. In the future, scientists will undoubtedly be able to provide a much better answer to the one question that comes to our minds when we stare into the eyes of a dolphin: What is he or she thinking about?

Chapter 5 Where Humans and Dolphins Meet

Baby Beluga in the deep blue sea,
Swim so wild and you swim so free.
Heaven above and the sea below,
And a little white whale on the go.
You're just a little white whale on the go.
RAFFI AND DEBI PIKE, "Baby Beluga"

It was like a fairy tale: the story of a beluga whale who befriended the humans of a small, rural seaside community in Nova Scotia. The beluga was young, not even three years old when first spotted by people. Probably she had recently lost her mother. A rumor circulated that a dead adult beluga with a bullet hole had washed ashore before the youngster appeared. This "baby" beluga seemed eager for the companionship of others . . . even if they were only humans. Wilma, as the locals called her, approached people on watercraft, whether small dinghies, small or large fishing vessels, or recreational boats. She often spyhopped (raised her head out of the water) and gradually allowed (and later encouraged!) people to touch the smooth, grayish white skin along her head and back. Perhaps most of all, children delighted in seeing her, and they sometimes came out in boats by the dozen. They lined up along the side of the boat as Wilma approached; with little hands, they reached over the gunwales, stretching their quivering fingers toward the water in hopes that this magical creature would swim near enough to allow them a chance to touch her. As the years progressed, and on days that were warm enough, people entered the water to swim with Wilma. She often accompanied swimmers and, if they didn't touch her, sometimes touched them with her flippers, often to the swimmers' surprise! This is how I came to meet a wild beluga and how my colleague Cathy Kinsman and I came to know her over the almost seven years that she interacted with this community. Our study of Wilma was the first ever conducted on a solitary, sociable beluga, and it was a unique opportunity for two very different species to learn about each other. Cathy and I observed many things about the behavior and communication of belugas that we would never have seen by studying them in larger groups in the wild or in captivity. This was only one of the many places where we have seen science and mythology overlap in the study of dolphin-human interactions.—Toni

The history of dolphin-human interactions is long-standing, complex, and unique in the animal world. Even without the enticement of food, it is not unusual for dolphins to initiate social contact with humans. Few wild animals are featured so prominently, or so positively, in human mythology, art, architecture, literature, anthropology, history, and the media. Somewhat paradoxically, dolphins are also treated as significant economic resources to humans, for recreation, education, tourism, entertainment, and sometimes as food or even fertilizer.[1]

Cultures around the world tell the story of our long interaction with dolphins in mythology, folklore, and real-life accounts. Petroglyphs of dolphins date back as far as nine thousand years. One of the earliest confirmed artistic representations of dolphins was created by the ancient Minoans on the island of Crete in 3500 B.C.E. Aesop's fable *The Monkey and the Dolphin* dates from the seventh century B.C.E., and Aristotle and other Greek thinkers regarded dolphins as highly intelligent creatures. Dolphins were celebrated in Greek and later Roman art, coinage, and architecture. Elsewhere in the world, the Maori of New Zealand, the Aborigines of Australia, the native tribes of the Amazon Basin, and the indigenous peoples of the Pacific Northwest Coast, have long represented dolphins in their art. In our time, beginning in the late 1950s, television shows such as *Flipper* and movies like *The Day of the Dolphin* all contributed to and reflected the elevated status of dolphins in popular culture.

The intense influence of human activity on dolphins and their habitat, combined with the unusually gregarious nature of dolphins toward people, has likely resulted

Two children interact with a beluga—sometimes
it seems the curiosity is mutual between cetaceans
and humans.

in interspecies cultural exchange and coevolution in some coastal dolphin and human societies over thousands of years. And yet, both the myth and the reality of dolphin-human interactions over the years are marred by a tragic irony: dolphins are highly vulnerable to negative impacts from the same human interactions that seek to celebrate them. How can dolphin-human communication alleviate this situation?

The father of modern-day study into dolphin-human communication was dolphin communication researcher John Lilly. Over more than eighty years of life Lilly held many titles: medical doctor, researcher at the National Institutes of Health, biophysicist, neuroscientist, cetacean scientist, writer, and inventor. He wrote no fewer than a dozen books, among them the classic *Man and Dolphin* (1961), and inspired several movies, including *The Day of the Dolphin* (1973) and *Altered States* (1980). Lilly was perhaps the first to explore, and then popularize, cetacean intelligence and self-awareness. In the late 1950s and 1960s, his early studies focused on neuroanatomy aimed at understanding intelligence. To do this research, he inserted electrodes into dolphin brains to study dolphins' vocal and mental abilities in a captive setting. This initial scientific approach was followed by drastically unconventional methods. Lilly became a member of a generation of counterculture scientists, thinkers, and artists. In one study he placed a lone dolphin in captivity with a woman who attempted to "live" with the dolphin. (This approach was somewhat reminiscent of experiments in which some early twentieth-century primatologists reared chimpanzees in human households to study the development of their behavior and communication.) Lilly eventually concluded that dolphins were as intelligent as humans, described dolphins as "extraterrestrial intelligence on earth," and espoused prophecies of communication between dolphins and humans that exceeded his data. His later work was largely dismissed by academics, but his prophecies infected popular culture regarding dolphins and interspecies communication.[2] Lilly's last published writing on dolphins, *Toward a Cetacean Nation*, summarizes his final messages about cetacean intelligence, the importance of protecting cetaceans, and giving them "rights as individuals under our laws."[3]

Lilly's work was one initial facet of a larger human movement to communicate with dolphins. Artists and writers such as Jim Nollman, who plays musical instruments with cetaceans and other animals around the world, and Wade Doak, who has ventured into many oceans to cavort with dolphins, have played an important role in the perception of how humans might be able to communicate with dolphins and what that connection might mean in human society.[4] Peter Beamish has described an innovative method of communicating with cetaceans via "rhythmic-," "signal-" based communication.[5] Others have attempted to communicate with wild and captive dolphins less systematically through music, computers, and other media.[6] Although we are learning a lot, we have not yet identified the key to dolphin-human communication through a shared language.

John Lilly's controversial methods of exploring dolphin-human communication led academia to dismiss this field for several decades. Over the years, however,

Dolphin and whale watching has become a
multibillion-dollar industry worldwide. But are
we loving these animals too much?

scientists slowly began to apply rigorous, systematic methods of inquiry and launched a new era of research on interspecies communication and dolphin-human interactions in particular.[7]

With guidance from my graduate adviser, Jane Packard, I (Toni) was apparently the first researcher to apply ethological techniques in assessing human-dolphin interactions and communication in both captive and wild settings.[8] In the late 1980s formalized and commercial swim programs with both captive and free-ranging dolphins were growing in popularity, and I believed there was a need to understand the dolphins' experience of these interactions and to determine to what degree, if any, interspecies communication was occurring.

In establishing a methodology for studying dolphin behavior and human-dolphin interactions, I looked at existing research on the effects of human activity, such as seismic exploration and boating, on baleen whales and dolphins. Yet this research focused primarily on such disturbances as boat activity and tuna-fishing operations rather than interpersonal, interspecies interactions.[9] In addition, these studies typically focused on the behavior of the group rather than the individual.[10]

To address my questions, I realized that I would need to meticulously observe and analyze the behaviors of individual dolphins ranging from subtle to overt actions. From video data, I documented behaviors in terms of their physical form rather than their functional significance, which allowed me to make reliable comparisons with other studies and avoid interpretive bias.[11] I evaluated the interaction

Dolphins seem to prefer to be the ones to approach swimmers. Paralleling their movements will yield a longer and more mutually enjoyable shared interaction.

and degree of interspecies communication that was occurring, and I found that the study of dolphin-human communication is readily comparable to the study of how different primate species interact. In fact, many techniques used to study human-human interactions, such as mother-child communication, can be applied to studying interspecies communication. This type of research has been found to be the most useful in studying the impacts of human activities, recreational or otherwise, on dolphins. Today a growing number of tremendously talented and diverse researchers specialize in the scientific assessment of the impacts of human activity on both toothed and baleen whales.[12] Through years of studying dolphin-human interactions with a focus on dolphins, we have found a portal through which to glimpse the emotions of dolphins as well as their communicative and cognitive processes—all of which enables us to better contribute to their welfare and conservation. The more I gaze through this portal, the more I learn how to focus my lens and improve my vision.

Do the behaviors that dolphins direct spontaneously toward humans constitute communication? As with *intra*species communication, qualifying *inter*species interactions as communication relies on the definitions and criteria chosen. Communication often occurs between members of a variety of species on many levels and, as we saw in chapter 2, is identified as a change in the behavior of the receiver in response to information provided by the signals of the sender.[13] Individuals may learn the referents and meanings of signals exhibited by members of another species and may even develop mutually understandable signs. In this way, communication between dolphins and humans may take place as it does between members of various dolphin species on different levels.[14]

Because the success of communication depends on the interaction between a sender and a receiver, individuals typically send signals that are easily read by, and are appropriately responded to, members of their own group, if not species.[15] Receivers thus often know what to expect from senders of the same species, and some level of predictability is achieved. Signals that have become specialized through evolutionary processes to communicate with peers are unlikely to be highly effective for communication between very dissimilar species.[16] In the first documented case of

Trainers communicate with belugas and dolphins every day. They share information through gesture, posture, and some pure-tone sounds.

dolphins interacting with another species —herbivorous dugongs and three species of tropical dolphins in the western Indian Ocean— the researchers proposed that "dugongs and dolphins were engaged in similar activities, such as traveling, on several occasions and were clearly associated when group formation was tight. If dolphins and dugongs may not associate for feeding purposes, then these interactions may occur: (1) for predation avoidance toward sharks; or (2) without any ecological reasons due to dolphin and dugong habitat overlap around the island."[17]

I (Toni) began my career in part because of my fascination with interspecies communication. Interspecies communication can occur, writes F. R. Walther, when one animal "acts and reacts toward the other as if it were a conspecific."[18] The zoologist Heini Hediger described this behavior as an "assimilation tendency": "A characteristic of men as well as animals . . . to regard animals of different species

. . . as if they belonged to the same species."[19] More recently Karen Pryor has noted that captive dolphins often respond to people as if they are dolphins.[20] In studying the interactions between people with dolphins in captivity and in the wild, I have found the most surprising results not in human-dolphin communication but rather in what I describe as *mis*communication, especially on the part of people who clearly misread dolphin signals. Although dolphins often alter their behaviors to accommodate people, people are generally less adept at reading and responding appropriately, at least in terms of dolphin etiquette. This is not just my interpretation of a dolphin code of ethics based on studying *intra*specific behavior. Dolphins often make themselves clear, if not forcefully or outright aggressively, in their response to people who are misbehaving in their world. Unfortunately, many people expect that dolphins should behave like Flipper or Shamu. They think that dolphins are perpetually cute, friendly, and thrilled to be in their presence. I have observed many situations in which dolphins exhibited warnings, or signs of frustration and stress, to people only to see those same people either ignoring the signs or interpreting them as an indication of pleasure. This has sometimes resulted in physical aggression by the dolphin. The title of Kathleen's chapter in the book *Between Species: Celebrating the Dolphin-Human Bond* is "Letting Dolphins Speak—Are We Listening?"[21] Clearly, sometimes, we are not. The biologists John Krebs and Richard Dawkins have stated that miscommunication can occur if a response is neutral or negative for a sender; during these situations, the responses are not good for either participant.[22]

The lesson of listening to, or astutely observing, the behavior of another animal is exemplified by our experiences with Pita, a free-ranging, lone, sociable bottlenose dolphin in Belize.[23] Pita was considered lone or solitary because she was rarely, if ever, in the company of others of her species while Kathleen and I studied her. Yet she was sociable and friendly with swimmers and boaters. Like many other solitary dolphins, Pita displayed a sense of possessiveness about certain objects in her environment. She frequented the sea around Lighthouse Reef Atoll and was visited almost daily by tourists from dive boats, sightseeing vessels, and fishermen. Pita's frequent and close interaction with snorkelers was replete with incidents during which she exhibited behavior that was more appropriate as a response to other dolphins. (The average snorkeler is no match for a free-ranging bottlenose dolphin.) Pita was particularly fond of anchor ropes, and she did not like to share. In one case, a snorkeler was holding the line. Pita swam up to her and nodded her head vertically, abruptly, and repeatedly, about 10 feet (3 m) in front of the snorkeler. The woman held on and appeared to exaggerate the display of her grasp by shaking the rope up and down underwater. Pita threatened to charge the woman and repeated her head nod. When the woman persisted, Pita displaced her by slowly but assertively swimming into her space, forcing her to let go of the anchor rope. After leaving the water, the woman stated that at first she thought Pita was playing with her and didn't realize until later that Pita might have assaulted her if she had not released her hold. Even though Pita won control of her "prize," she "hung" stationary in the water with her flukes hanging down as if they were heavy with the weight of the anchor itself. If I ever saw a dolphin "pout," this would be it. After seeing numerous examples in which people viewed dolphins through "Flipper-colored glasses" and the less-than-ideal consequences that followed these interactions, I reluctantly reoriented my specialization to the study of interspecies *mis*communication—a less glamorous, but I hope helpful, contribution to our understanding of what dolphins are attempting to communicate with us that we are not perceiving in the way that another dolphin might.

Mimicry of human voices, human-associated sounds, and human postures have often been observed from both captive and free-ranging dolphins.[24] Mimicry can be a form of play, a method of communication, or both. When in the water with dolphins I (Toni) have often observed them imitating my behavior and the actions of others. Once I was swimming underwater and filming spinner dolphins when a colleague began slapping his hand repeatedly on the water's surface. He was trying to get my attention. When I looked up, I saw a juvenile dolphin, about 30 feet (5 m) away from my colleague and out of his view. She was also at the surface and was slapping her pectoral fin against the water, apparently in mimicry of him. After I pointed to the dolphin, he noticed her. The context was playful, and because she remained near us for at least ten minutes more, we believed that her action was one of mimicry. Of course, she might have been signaling to another dolphin out of our view, which

Dolphins will often engage our imaginations whether we observe them firsthand underwater or through the media.

could suggest a different message.

People who work with captive dolphins sometimes find their own voices mimicked by dolphins. Lilly expounded on dolphins' propensity to imitate human sounds in his experiments by training dolphins to say certain words. Does anyone remember the dolphins saying "Ma," "Pa," and "Fa" in the science-fiction thriller *The Day of the Dolphin?* Still, Lilly concluded that because the structure from which the dolphin emits sound is so different from the vocal cords of humans, people would do best to work with dolphins in a more suitable, dolphinlike manner. That is, we should imitate their sounds and not ask them to imitate human speech.

In each of the solitary, sociable individuals that I have studied (bottlenose dolphins, killer whales, and belugas), I have observed occurrences of mimicry of human behavior or human-made sounds, especially if the individual had been in the company of humans for extended periods. Other researchers have also noted this behavior.[25] I have heard these animals make sounds that were distinctive impersonations of boat motors. In some cases, the dolphin or beluga mimic actually rubbed against or touched an outboard motor while making the strangest sounding motor noises I have ever heard!

It appears that dolphins who frequently interact with humans in the wild have learned how to communicate with humans on some level, perhaps to solicit certain behavioral responses from swimmers. For example, free-ranging spotted dolphins, a solitary bottlenose dolphin, and captive dolphins have all exhibited postural mimicry of swimmers.[26] In 1992, about five days after Hurricane Andrew passed over The Bahamas, we (Kathleen and three ecotourists) encountered four juvenile and subadult spotted dolphins not more than a mile or two (a couple of kilometers) from West End. Just after the hurricane, the dolphins wanted nothing to do with us: resting, foraging, and "regrouping" after the storm seemed their priority. A couple of days later, however, it was a different story. We spent two and a half hours with this foursome. I could regale you with many stories, but one observation sticks out in my mind. Off to my right, I watched as an ecotourist named Jennifer swam tight circles with ID#65, a subadult female we nicknamed Venus. At first, their circles were wide, maybe 6 feet (2 m) in diameter, but as these two individuals swam faster, their circles got tighter until they collided! Body-to-body they clobbered each other, stopped to shake it off, looked at each other, and then resumed their circling mimicry. This continued for about twenty minutes as I filmed the action (my role is the objective observer). After the encounter, Jennifer was talking a mile a minute and was jazzed by her interaction. Jennifer felt that Venus left with the same enthusiasm that she did. I often wonder who started that game, but Jennifer couldn't say. This is not a question I get to ask very often because my studies focus on dolphin-dolphin behavior. Still, sometimes watching a game like Jennifer's sheds light on why dolphins might initiate a similar game with themselves: sheer joy is one explanation, and the smile on Jennifer's face certainly supported that hypothesis.

Similarly, when dolphins interact with human swimmers, at times they seem to modify their actions to accommodate their human associates, even without intentional reinforcement.[27] I (Toni) have seen one or more dolphins in the wild separate from their group to remain closer to the surface, swim more slowly, and stay near certain swimmers. Above the surface, both in captivity and in the wild, dolphins often modify the production and quality of their vocalizations. Cathy Kinsman of the Whale Stewardship Project and I have observed numerous solitary, sociable belugas, orcas, and bottlenose dolphins emitting vocalizations above the water instead of underwater when interacting with people from boats. We have also seen this happen spontaneously at poolside with captive dolphins—such as with trainers or in public feed-the-dolphin programs. Of course, although we are not training the dolphins to do this, we could be unintentionally reinforcing such behaviors simply by responding to the dolphins when they spontaneously accommodate us.

Another great example of interspecies communication is play behavior, especially between dolphins and humans who enjoy playing in their daily lives.[28] Conversely, play can serve as an important mechanism in the development of communication skills.[29] Play between dolphins and humans might provide the motivation for the development of mutually understood signals needed for us to communicate with each other.

We have documented instances of what Kathleen calls "rebellious" play and what

Dolphins will often
mimic or mirror the
actions of swimmers,
which can lead to tight
circles or circle swims.

Four teenage Indo-Pacific bottlenose dolphins use an octopus as their ball for a game of catch. Dolphins often play with other sea creatures whether they are willing participants or not!

Toni calls "mischievous" play. Call it what you will . . . dolphins can play dirty. Late one afternoon around Mikura Island, Kathleen watched as one dolphin pulled an octopus from a rock crevice. The dolphin and three of his friends then used that octopus as a ball! They seemed to delight in their short-lived game, but the octopus had the last laugh: as the four dolphins sped from view, I saw that the octopus was actually suckered to the face of one of these pranksters. Over the years, we have both observed dolphins having a grand time playing with each other, mimicking human swimmers (sometimes in unflattering ways!), playfully terrorizing fish both small and large, and donning seaweed as if it were the crown jewels or a cashmere shawl, only to flaunt these objects or use them to play keep-away or tag with swimmers. Cathy Kinsman has seen a free-ranging, solitary beluga squirt water at someone in a boat. Toni worked with two captive dolphins that had an uncanny ability to aim water from their mouths directly at people and their cameras (not the waterproof version, mind you) before posing for a picture. The screams emitted by the would-be photographers must have truly entertained the dolphins, which seemed to have quite a twinkle in their eye on the expert firing of their squirt.

Dolphins' mischievous play with humans, at least as we interpret it, can practically wrap people into a ball. As part of my (Toni's) doctoral research on spotted and bottlenose dolphins in The Bahamas, I found my own mind spinning in circles while analyzing numerous hours of recorded video of dolphins circling swimmers.[30] Dolphins from both study populations were observed circling one another as well as humans.[31] I was so intent on understanding the components of this behavior that I dissected each element of the interactions. Alas, with all my numbers, nothing in my data stood out, and I was unable to glean any meaning from this game. Perhaps that is what I get for trying to make serious sense out of something that could well be just plain fun.

Though not considered common, play between species does occur.[32] As non-utilitarian interest in animals has increased, unexpected and amazing accounts of play between different animal species have been reported, such as one case in which a crow and kitten befriended each other and were often found playing together. These observations seem to occur most frequently when at least one individual in the cavorting pair is very young and especially if others of their species are unavailable. Of those species that do interact socially with others, such as dogs and humans, play may be a predominant form of interaction. As mentioned above, many people interacting with dolphins have found themselves partaking in games of keep-away and drop and chase with seaweed and other objects.[33] And such games are frequently observed within dolphin societies. Researchers Kenneth Norris and Thomas Dohl have suggested that play was a component of social interactions between free-ranging, solitary bottlenose dolphins and swimmers who frequently interacted with one another.[34] They identified behaviors such as dolphins allowing humans to ride or be "towed" as examples. One study of dolphins used in a captive swim program reported a variety of play behaviors exhibited by dolphins directed toward swimmers; it was found that play accounted for 39 percent of the dolphin-initiated interaction time with humans.[35]

Touch may also be a form of play, and some forms of touch in dolphins have been identified as sexual play.[36] With the exception of dolphins who are provisioned with food and solitary dolphins who frequently interact with swimmers, free-ranging dolphins rarely allow tactile contact with swimmers.[37] The surprising occurrence

Dolphins are known to tease and play with birds. Dolphins have thrown fish to birds, and a few killer whales have been seen baiting gulls that fly over their pools.

and variety of dolphin sexual behavior directed toward people provides insight into how sexual behavior serves as a form of social interaction. Intraspecific sexual behavior in dolphins of both genders and in virtually all age classes is frequently observed, heterosexually and homosexually. People working with captive as well as solitary, sociable dolphins have long known, and lamented, that male dolphins with erections will readily rub against any object in their environment, including other dolphins and human body parts (behind the knees are a favorite). Females also rub against humans, though not nearly as often as males. Although people might think this is a form of flattery or, conversely, a reason to be scared out of their wits, they should remember that sexual behavior in dolphins can serve different functions than it does in humans or other animal species. Still, there is reason to be cautious around amorous dolphins. One study of captive dolphins in swim programs showed a clear correlation between sexual and aggressive behavior directed toward humans.[38] And, researchers

Many trainers speak of having the dolphins with whom they work spending time watching or looking for them, waiting expectantly near the pool sides for the trainers to return. This behavior can be modified and shaped so that trainers can coordinate behavioral requests from animals through underwater viewing ports.

like myself (Toni), who have studied solitary dolphins, especially males, interacting with swimmers in the wild, have observed a similar association between these two actions. Male bottlenose dolphins use sexual behavior to establish dominance hierarchies, raising the question of whether they could be using this behavior to establish dominance over humans.

The training of dolphins (or perhaps, we often wonder, it is dolphins training the humans) is a venue for the in-depth exploration of dolphin-human communication. Dolphins in captivity and trainers communicate through positive reinforcement: dolphins learn to modify their behavior to obtain a reward according to signals learned from the trainer.[39] Conversely, trainers learn the significance of "intraspecific social signals" that dolphins direct toward them; the trainer then adapts to these signals to alter the dolphin's behavior.[40] Dolphin signals directed toward trainers are not always behaviors typically seen in the wild, with the exception of free-ranging dolphins that are given food.[41] (When we study communication between species, we use the term *signal* in the context of reinforcement training.[42] This is a different usage of *signal* than used in communication theory, in which signals are specialized species-specific behaviors.)[43] For example, when a trainer is on the dock or platform, captive dolphins often position themselves in front of the trainer with their heads out of the water for extended periods. Wild dolphins have been observed to spyhop in this fashion, though not typically for the lengthy periods as when watching trainers.

Most dolphin trainers will tell you that their work with dolphins focuses not only on stimulus-response episodes but also on establishing a social relationship with the dolphins in their care. They develop this relationship through communication, both overt and subtle.[44] Dolphins typically are extremely interested in humans, and their gregarious behavior is the basis for the trainer-dolphin relationship. During training, the trainer behaves consistently toward the dolphin, which gives the dolphin an opportunity to predict and, in turn, possibly to manipulate, the behavior of the trainer to some degree. Spontaneous dolphin behavior is another important aspect of reinforcement training; the trainer can encourage, positively reinforce, and shape the behavior.[45]

Communication that occurs between a dolphin and a trainer during captive swim-with-the-dolphin programs provides a social context in which dolphins interact with both familiar trainers and unfamiliar swimmers.[46] One might expect a dolphin to behave differently toward an experienced trainer with whom the dolphin has a relationship as opposed to an unfamiliar, naive swimmer.[47]

Given the differences in the environments and anatomical and morphological characteristics of dolphins and primates, it is interesting that so many aspects of their social behavior are similar.[48] For example, dolphins and primates both maintain close social bonds that can last for many years.[49] Both live in fusion-fission societies: females often rely on other females as allies, males form long-term coalitions that are significant in reproductive competition, and females often remain in core areas while males tend to disperse and range more widely.[50] Members of both groups have been observed to engage in cooperative hunting as well as mutual defense.[51]

Early studies of dolphins indicated that they learn more quickly through auditory than visual cues.[52] The opposite has been documented for primates.[53] More recent studies in dolphins found evidence of impressive visual processing in conjunction with echolocation.[54] The importance of visual and audio perception in learning and communication thus appears to be shared between primates and dolphins. They also share many similar aspects of communication: both dolphins and primates are highly social, with visual postures and gestures, tactile interactions, and vocalizations important aspects of their societies.[55] Morphology (for example, sexual dimorphism, coloration) is also important in both dolphin and primate communication.[56] Social spacing, positioning, and orientation may be related to the social rank of individual dolphins and primates: males of both groups form complex alliances.[57]

Play represents a large part of the behavioral repertoire of both primates and dolphins, is observed in high frequency in both groups, and is described similarly for both groups in terms of complexity and context.[58] The rhythmic stamping of chimpanzees as documented by Wolfgang Köhler in 1925 was later compared with the rhythmic tail slapping observed in dolphins. Play with inanimate objects, play as a mechanism for sexual learning, as well as play directed toward members of other species, are observed in both groups, as well as in other animals.[59]

The line between fact and fable is blurred by tales of human-dolphin interactions from classical antiquity to the present. Plutarch wrote of the dolphin in the first century c.e. that "it has no need of any man, yet is the friend of all men and has often given them great aid." Stories of free-ranging dolphins "befriending" humans and giving them rides at sea, saving people's lives, and fishing cooperatively with them continue from ancient times to the present. Associations between dolphins and humans typically develop when both experience repeated mutual reinforcement of positive sociable behavior: both enjoy the connection. (With a few exceptions, feeding dolphins has not been a typical component of sociable relations between dolphins and people.)[60]

The Roman naturalist Pliny the Elder recorded one of the earliest descriptions of a friendly solitary dolphin. As he told it, a dolphin named Simo in the Mediterranean formed a close relationship with a boy and even gave the boy "rides" to and from school. In modern times, a solitary, sociable bottlenose dolphin, who would play with boaters and bathers, was given the name Opo by the people of Opononi, New Zealand, in 1955. Since Opo, more than seventy solitary, sociable dolphins have been identified in coastal locations around the world.[61] These individuals may interact with people in boats, on shore, wading, or swimming with various degrees of contact and habituation.

Cooperative fishing is another example of interspecies communication that can benefit both dolphins and humans.[62] Apparently, dolphins, not fishermen, initiate these highly sophisticated relationships in which they work together to herd and catch fish; various species of river and oceanic dolphins have been involved. Pliny the Elder wrote that dolphins worked with humans to drive mullet into nets in the Rhone River in modern-day France. The Australian aborigines of Queensland's Moreton Bay and Amazonian fishermen have fished with regular assistance of dolphins. In Laguna, Brazil, fishermen continue to fish cooperatively with bottlenose dolphins as they have done for generations.[63] People work exclusively with one or several particular dolphins. This relationship sometimes extends to the progeny of both the dolphins and the fishermen for several generations. In the mid-1800s a well-known group of killer whales in New South Wales, Australia, engaged in cooperative hunting of whales with humans for at least five decades.[64] Cooperative fishing between dolphins and sea birds, such as pelicans, has also been observed with some frequency and may provide additional insight into the nature of human-dolphin fishing cooperatives.

What does it mean for a dolphin to be alone in the wild? Since most toothed whales are known for their sophisticated social structures and complex communication

It is extremely rare to see a dolphin consistently alone; dolphins are gregarious, tactile, social animals and typically are found in small to large groups. Lone, sociable dolphins have been observed since classical times, but this is not common.

skills, a lone dolphin is considered an anomaly. Yet with increasing frequency we are seeing greater numbers of solitary, sociable toothed whales. They are solitary because they are rarely if ever observed with others of their kind. They are sociable because they initiate outgoing, gregarious interactions with humans including boaters, swimmers, people standing at the shore, or a combination of all of these. Not all solitary individuals that are normally seen in groups are immediately social with people. Still, it seems only a matter of time for each of these lone rangers to develop a relationship with people, especially if they remain apart from members of their own species and near relatively nonthreatening humans for enough time.

Lone, sociable dolphins are now recognized worldwide, with some individuals studied more intensively than others.[65] Both of us studied Pita, the lone bottlenose that called Lighthouse Reef Atoll, Belize, home for more than fifteen years.[66] Although most accounts of lone, sociable dolphins are coastal bottlenose dolphins of both genders, this trend has shifted over the past decade. My (Toni's) research has followed suit. Cathy Kinsman invited me to conduct the first scientific studies on lone, sociable beluga whales, and in 1998 we began studying Wilma, the first of six solitary belugas we followed closely over the next decade. When I was asked about my predictions for this first solitary, sociable beluga, I projected that she would exhibit many of the same behaviors and experience similar outcomes as many of the solitary, sociable bottlenose dolphins. Unfortunately, my forecast for Wilma and for most of the solitary belugas we studied was true. Solitary, sociable dolphins are uniquely vulnerable to injury and death—whether intentionally or unintentionally—by human hands or boat propellers. With each solitary individual I hear of, I hope that humans learn respect and tolerance and that the solitaries survive.

A few years after Wilma arrived on the scene, two young solitary killer whales from local resident pods were observed completely alone in the northeastern Pacific —practically in my aquatic backyard of Puget Sound. In 2001, Luna (L98), a two-and-a-half-year-old male, was seen in Nootka Sound, British Columbia. He spent time in an area within the geographic range of southern resident orcas but remained isolated within an inlet away from the presumed travel routes of this population. In January 2002, in waters around Vashon Island, Washington, an area outside the usual travel routes of the northern resident orcas, boaters and researchers sighted a female orca estimated at roughly two years old and named her Springer (A73). She was an orphan from the northern resident group and did not associate with members of her natal pod. It is important to keep in mind how uniquely and intensely

Wilma showed interest in boat propellers and even expertly mimicked the noises made by boat motors.

bonded the resident orcas of the Pacific Northwest are and how both daughters and sons typically remain with their biological pods for life. We saw Luna first, but Springer, possibly because she resided in more urban waters, received more attention from the public and the media. When the media and orca experts (of which I am not one) queried me for predictions regarding the potential fate of each of these animals, I referred to their persistent inquisitive behavior and repeated solicitation of human interaction, which suggested that each would succumb to the typical fate of a solitary dolphin. I was wrong on one count; there were two very different conclusions, which reflect how each situation was managed. Springer successfully reunited with her natal pod, whereas Luna paid the ultimate price during an interaction with a propeller.

"Why are these animals solitary?" is the question I hear most often. It is also perhaps the hardest to answer. The reasons vary greatly, from knowing that the animal was orphaned at an early age to knowing nothing about the animal's history before his or her appearance alone. Although the situation is unfortunate for the individuals, these sociable odontocetes provide incredible opportunities for researchers to study the development of cognition, communication, and social behavior. These solitaries present something like a cross between captive dolphins and free-ranging dolphins in social groups; they exist in a socially impoverished environment with respect to peers, yet their habitat is relatively rich and natural. Designing "enrichment programs" for solitary odontocetes is similar to creating the same for captive dolphins; the programs may differ in the items used for enriching the quality of their lives, but the intent is the same.

In the past few decades in many parts of the world, it has become far more popular and profitable to watch whales and dolphins than to kill them.[67] In part, this is because of the friendly, outgoing nature of most dolphins. People can view or intermingle with dolphins in activities ranging from nonconsumptive (watching them from land) to minimally consumptive (watching them from boats that are operated knowledgeably and cautiously) to highly consumptive (overt harassment or capture from the wild).[68] The appeal of swimming, or otherwise closely connecting, with dolphins both in captivity and in the wild has been commercialized into

Kathleen exits the water after a session recording dolphin behavior and sounds from among bottlenose dolphins at the Roatan Institute for Marine Sciences, Roatan, Honduras. Gracie, an adult female dolphin, watches her exit, while Robin Paulos, one of Kathleen's doctoral students, assists Kathleen.

an international, multibillion-dollar industry in which dolphins are often exploited. Dolphins are increasingly sought after as lucrative sources of entertainment and recreation, as well as, to a lesser degree, for educational or therapeutic purposes. Responsible research, management, and public awareness of dolphin interactions have lagged sorely behind this expansion, resulting in dangerous and occasionally fatal consequences for the dolphins. This is not to say there are not responsible operations in existence—there are. But scientific research examining the interactions and effect of humans on dolphins has not been high on people's list of priorities.

We both study wild dolphins; we both study captive dolphins. And we began our studies on captive dolphins with open minds and mostly similar views. Over time, our experiences with diverse facilities and our different research on dolphins led us to divergent perspectives regarding the countless issues involving captive dolphins. We agree on several major points, such as on having well-defined and enforceable legal requirements for the maintenance of dolphins in captivity and strict management policies on interactions with wild dolphins.

In the early 1990s, I (Toni) conducted a study on the behavior of dolphins in a captive swim program. Since then, I have inspected numerous captive facilities at the request of government and nongovernment agencies worldwide and have often investigated facilities that received complaints by visitors or agency personnel. Not surprisingly, many of these requests brought me to some of the most poorly managed sites in the world. For this reason, much of my experience with captive display facilities has been very different from that of Kathleen's.

One particularly grim experience was when two veterinarians and I were hired by The Bahamian government in 1994 to inspect three captive swim-with-the-dolphin facilities. The conditions at one location were so disturbing that all three of us recommended it be closed, which the government endorsed.[69] This place had obtained several dolphins without a permit, and the Department of Fisheries (now the Department of Marine Resources) requested that the owner release these dolphins into the wild. One male had been kept in isolation because he exhibited overly sexual and aggressive behavior toward participants in the swim program. During our visit, he incessantly swam in circles, exhibiting what is known as stereotypic behavior, which is often related to stress. The dolphin that had previously lived in his pool had reportedly died (before our arrival) of suffocation after entanglement in a submerged rope. Unfortunately, I have observed too many instances of excessively poor conditions for captive dolphins in countries ranging from developing nations to the United States and western European countries.

In 1994, I (Kathleen) began my studies of captive dolphins with a month of observations at Kolmården Djurpark in Sweden. I was there to test the accuracy and precision of my mobile video/acoustic system against a stationary array of hydrophones. It would be nine years before I could launch a full-scale study of a group of captive dolphins, but those first few weeks of observation remain strong in my memory. It took about a week for the dolphins to habituate to my presence and to accept that I would not interact with them. I never touched the dolphins at Kolmården nor interacted with them; I was privileged, however, to observe both juvenile females practice behaviors that the trainers had not yet taught them. The younger of these two females had a game she often played: she beached on the floating dock and then rolled onto an adult as he or she swam by. She delighted in this game, and did it often until she rolled onto the adult male. He did not discipline the young dolphin but went directly to her mom, who in turn disciplined her calf. My experiences observing dolphins in captivity have presented to me healthy sociability among the groups. I have not witnessed major differences between the wild dolphins that I study with the three groups of captive dolphins I have observed. As Toni has rightly mentioned, there are hundreds of facilities worldwide, and not all have the well-being of the animals in their care as a priority.

An unfortunate trend for dolphins is their increasing use in "therapeutic" settings by some psychologists and spiritualists. Programs using both wild and captive dolphins for human therapy sessions are growing exponentially in number. Independent researchers conclude that claims about dolphin-assisted therapy for people with illnesses or disabilities are not substantiated by valid research and are not ethically or professionally justifiable.[70] The data indicate that similar, if not better, results are often obtained using domestic animals or other therapeutic methods such as hydrotherapy.[71] One study also noted that "many patients hesitated to interact with the dolphins in the sessions because they were scared by these huge, unknown animals."[72] Even the founder of dolphin-assisted therapy now recommends against the practice, concluding that "people would never throw their child in with a strange dog, but they'll throw them in with a strange dolphin. What you are looking at are vulnerable people and vulnerable dolphins."[73] Some of the controversy regarding these programs originates from people who spend large sums of money only to find that claims made regarding dolphin cures do not exist.

Another factor to ponder when considering dolphin therapy is whether potential confounding elements of the experience can be separated. Is there a specific direct action made by a dolphin that heals a person? Do dolphins have special echolocation that can be directed at harmful cells in a person's body? Does a chemical reaction or other neural change occur within the brain of the patient when he or she is in the water near a dolphin? Many studies have been conducted to understand the therapeutic effects of dolphins on a variety of human ailments.[74] The results are mildly conflicting if not outright contradictory. We do not yet know enough about the human brain to determine whether a nonhuman animal species, whether dolphin, horse, or dog, can have a direct healing effect on a human ailment.

It may be hard to believe that people and dolphins can share the same illnesses, considering our different physical appearances and our different living environments. Yet we are all mammals, breathe the same air when together, and share many biological traits. Disease transmission is in fact a serious concern, because dolphins carry maladies that can be transmitted to humans, and vice versa.[75] In a report to the U.S. Marine Mammal Commission of people who interact physically with marine mammals, primarily trainers and rehabilitators, 23 percent of respondents reported a physical ailment that was believed to be related to their contact with the animals.[76] Respiratory diseases such as tuberculosis were reported in roughly a fifth of marine mammal workers.[77]

Regardless of the realities and risks encountered when people swim with dolphins, we are amazed at the ways these animals occasionally live up to their mythical reputations. People of the Amazon Basin tell of river dolphins saving the lives of villagers whose boats capsized. There are numerous accounts of dolphins assisting swimmers in distress who otherwise would have drowned; dolphins rendered aid by holding the swimmer at the surface, pushing them to shore, or guiding their boats to shore in the fog, in a storm, or in the dark.[78] There are even reports of dolphins protecting swimmers from sharks.[79] One highly publicized account was that of Elián González, the little boy refugee from Cuba, who said he was "saved by dolphins" when he was close to drowning. The dolphins held him above the water by supporting his body with their bodies; Elián also reported that the dolphins chased away sharks.[80]

Another poignant story shows the fragility of dolphin-human contacts. On November 24, 2004, the *New Zealand Herald* reported the story of a veteran lifeguard who credited a group of bottlenose dolphins with saving four people—the lifeguard and three teenagers—from a circling great white shark for more than half an hour. "They were behaving really weird," said the lifeguard, ". . . turning tight circles on us and slapping the water with their tails. . . . The dolphins were going ballistic."[81] Many witnesses corroborated the story, which captured the imagination and hearts of people around the world. It was featured on CNN International and the front page of the London *Times* and became the basis for a docudrama for the BBC's *Natural World* series. Sadly, two days later, the *New Zealand Herald* featured a very different story. Illegal net fishing was now blamed for the deaths of two dolphins, which had their tails cut off presumably to release their bodies from the nets. "This is how we repay them for their help?!" the lifeguard angrily asked.

We have both seen dolphins assist people in distress ourselves. When I (Toni) was doing research from a boat offshore in The Bahamas, about eight swimmers and I were in the water with the dolphins. Each of us was swimming with a different group or subgroup of dolphins, and I was videotaping the behavior of the dolphins closest to me. In a flash, the dolphins disappeared from my viewfinder, and when I looked for them I saw they had joined a larger group that had suddenly formed. A woman from our group, at the surface, was in the center of all the dolphins. When I looked more closely, I realized she was barely swimming and was, in fact, struggling in the water. By then several swimmers had also joined her and learned that she had become suddenly seriously fatigued. She later told us that just as she became afraid and was about to call for help, the dolphins came from all directions to surround her. Two of the men in our group helped her back to the boat, and the dolphins swam off to do whatever dolphins do after such an event.

Dolphins do not always assist people in danger, and on occasion they may even harm swimmers.[82] In 2006, I (Toni) served as a visiting scientist for the National Geographic television series *Hunter and the Hunted*. The episode was called "Dolphin Attack," which aptly describes the footage they asked me to submit and describe to viewers. I scrutinized roughly twenty segments of incidents in which various species of dolphin visibly attacked humans. I hesitated at first about participating in a show that would portray dolphins as vicious animals that attack without provocation. Ultimately, I decided it was best to participate to help dispel the "happy-go-lucky, friendly helper to man" myth of the dolphin. And, despite the adrenaline-boosting previews of the show, the producers were true to their word that they would include a relatively thorough discussion of the reasons behind such attacks.

Incidents in which dolphins harm humans are typically limited to specific abnormal or stressful conditions, such as in the absence of companionship of other dolphins, in situations where the public regularly feeds them, when intentionally or unintentionally disturbed or harassed, or at times in captivity. These episodes are occurring more frequently with the increasing public demand for close interactions with these wild animals. Dolphins are not always friendly and peaceful with members of their *own* species. Infanticide has been witnessed among bottlenose dolphins in several locations around the globe.[83] Coalitions of male bottlenose dolphins will coordinate their activity to herd females for mating; their actions and aggression often cause severe lacerations to the target female, if not to sparring males. Home range territoriality or dolphins' lack of control over people entering their habitat may also affect their behavior. Dolphins naturally have a large habitat.[84] Large ranges permit dolphins to be apart, preventing aggressive encounters.[85] The natural environment allows for dolphins to disperse from one another during social tensions, reducing the level of stress from direct conflict.[86]

Directed aggression toward humans by dolphins who interact with humans frequently may be related to their expecting humans to behave as if they were other dolphins.[87] If a person violates the "rules" of delphinid social interaction, a dolphin that has acclimated to humans may direct the same dominance-related behavior toward humans that is shared with peers.[88] The dolphin, in effect, may also expect the swimmer to fill the social roles normally occupied by peers. Dolphins are certainly capable of discriminating between humans and other dolphins and can respond to members of each species, as well as to individuals, accordingly. Still, they will direct aggressive and other behaviors toward humans as they would peers—perhaps as some people do with other animals, such as their dogs.

The only confirmed account of a human fatality resulting from a bottlenose dolphin occurred in Brazil when a solitary, sociable dolphin was physically restrained and abused by two inebriated men, an obvious case of self-defense as documented by biologist Marcos Santos.[89] Two other human fatalities (one involving the animals' trainer) and many more injuries have been attributed to captive killer whales.[90] It is not uncommon for members of the public to become injured from swimming with captive dolphins or even from interacting with them at poolside; injuries have included broken bones, internal injuries, bruises, and lacerations, some requiring hospitalization. Even trainers with extensive experience with the dolphins in their care have been seriously injured.

In one study, 52 percent of people working with marine mammals reported marine mammal–inflicted injuries, and more than a third of these injuries were considered severe.[91] The only time I (Toni) was ever truly afraid of dolphins was when I was in the water conducting preliminary observations for a study on captive swim programs. Two males with obvious erections were competing for access to a female in the enclosure. The female attempted to interact with me (perhaps as a distraction from the males?). The next thing I knew, all three of them were tumbling over and around one another, with me in the middle. As I tried to swim away toward the platform to get out, the tumbling became rougher. Given my predicament, I made a more serious move and finally exited the water, fortunate not to have been bruised, let alone broken, in the dolphin drama that continued.

One of my (Kathleen's) early research collaborations focused on studying Pita in Belize. I video-recorded her off of Lighthouse Reef Atoll to determine her gender and to document her identifying scars and marks. I will always remember my first, and only, swim with Pita. She was sighted from the boat, we stopped, and I entered the water with my camera. I had relatively good visibility, but I could not see or hear her. Then someone from the boat told me to turn around; I did and was literally face to face with Pita. She seemed huge and intimidating. After about fifteen minutes of video, Pita positioned her body parallel to my body but between the water's surface and me. We were in about 4 to 5 feet (1.5 m) of water and Pita was intentionally in my space. Still, after almost two dozen years of studying wild and captive dolphin behavior, having been jaw clapped at and pushed and prodded, this is the only time I've ever felt uneasy in the water with a dolphin. I pay attention to dolphins and work very hard at being noninvasive. Both Toni and

I may be two of the only people on the planet who would prefer that dolphins ignored us.

All the records of dolphins saving people, from the ancient Greeks to today, represent a biased sample; that is, we typically never hear from the people whom dolphins do not save. We would be interested to know what criteria dolphins employ when deciding whom to save and whom not to save. We caution anyone against depending on dolphins to save their lives if in danger at sea. In The Bahamas, Toni was once abandoned by dolphins she was observing underwater, only to find herself swimming with a shark instead! And we have received infrequent but reliable reports of pilot whales, killer whales, and other cetaceans intentionally striking small boats until, on rare occasions, they have been seriously damaged. When these behaviors are considered, it appears that people can have a considerable range of experiences while interacting with dolphins, from becoming injured to being rescued. Although this does not help perpetuate a mystical stereotype, it does promote a more complex and intriguing portrait of these animals. This complexity of dolphins, as well as the individuality that they exhibit when we look at them closely, is what we believe makes the facts about them even more fascinating than the myths.

Chapter 6 Communicating Conservation

We need another and a wiser and perhaps a more mystical concept of animals. Remote from universal nature, and living by complicated artifice, man in civilization surveys the creature through the glass of his knowledge and sees thereby a feather magnified and the whole image distorted. We patronize them for their incompleteness, for their tragic fate of having taken form so far below ourselves. And therein we err, and greatly err. For the animal shall not be measured by man. In a world older and more complex than ours they move finished and complete, gifted with extensions of the senses we have lost or never attained, living by voices we shall never hear. They are not brethren, they are not underlings; they are other nations, caught with ourselves in the net of life and time, fellow prisoners of the splendour and travail of the earth.

HENRY BESTON, *The Outermost House*

In San Ignacio Lagoon in Baja, Mexico, a gray whale closely approaches our little boat. Gray whales have the longest annual migration of any mammal; they mate and give birth to their calves in this lagoon and forage way north in the frigid waters off Alaska. I am here to observe the impact that proposed salt mines might have on these whales. Unlike anything I've ever witnessed before, these friendly whales keep me mesmerized. Our close visitor is a mom, and her shiny new calf is barely the size of our boat. The mom lolls to one side to see us better— she gazes at us while the calf swims curiously around her. Mom is practically touching our boat when she slowly lifts her head above the surface to receive our outstretched hands with surprisingly gentle and soft caresses. I surprise myself by participating in the "whale massage." I found this situation to be an exception to my otherwise rigid rule of "no touching wild animals." This mother whale bears a scar resembling an old harpoon injury—a glaring reminder that her species was almost hunted to extinction before an international moratorium on commercial whaling was placed along their Pacific migration route in the 1940s. But this quiet moment with mom and her calf represented a unique interspecies trust upon which an entire ecotourism and international public conservation program was created. After this trip, I learned that the proposal to build the salt mines had been defeated; the decision was strongly tied to the ecotourism industry developing around these whales. It took one of the largest cetaceans to help me realize how powerful communication between humans and animals can be for the conservation of a species.—Toni

The study of dolphin communication directly contributes to our ability to protect and enhance the welfare and conservation of cetaceans. It also guides us in assessing the effectiveness of such efforts. My (Kathleen's) research has been cosponsored by ecotourism since 1991: people interested in learning more about dolphins or those individuals who just want to meet dolphins up close and personal join me during some portions of my fieldwork for data collection. I remember a trip to Belize in 1992 with a group of eight participants. Aboard our survey boat about midway through the trip, one of the women turned to me with an astonished look. She said, "I am awed by how much we do not know about dolphins!" She had come on the trip to learn everything that could be known about dolphins but realized that there is still much (even now, many years later) to learn about these aquatic mammals. If we can reach one or two people every month with a sentiment like hers, then we will be well on our way to fostering a community of committed environmental stewards.

The effects of human activity on the behavior of dolphins can harm their survival and reproductive success.[1] A variety of behaviors like avoidance, tail slapping, chuffing (like a "huff" from the blowhole), swift departure, charging, and ramming may indicate a negative reaction by dolphins to some human action.[2] Vessel traffic affects the behavior of many species of toothed whale, including killer whales, spotted, spinner, bottlenose, striped, and common dolphins, harbor porpoises, narwhals, and belugas.³ Observations both above and below the surface have been conducted to assess the effects of boat and swimmer activity on dolphin behavior. Above-water annotations revealed an apparent absence of stress-related behaviors in dolphins near swimmers. In New Zealand, Rochelle Constantine determined that dolphin behavior toward swimmers was somewhat dependent on the boat's operation and placement of the swimmers. She found that dolphins exhibited the least avoidance behavior when swimmers were placed abreast of the dolphins, while putting the swimmers in the water when the dolphins were bow-riding yielded the highest rate of sustained in-water interaction: the dolphins remained near the boat longer. Constantine also recorded the highest avoidance of swimmers when boats placed swimmers in their path of travel.⁴ We need to understand what dolphins consider acceptable before imposing ourselves on them. Understanding dolphin society and their information signals better informs our management of these interactions. We need to communicate our findings about dolphins to instill better manners in humans when interacting around dolphins. This includes learning how to avoid miscommunication.

Many dolphin species face serious wildlife conservation challenges today, and some species are on the verge of extinction. In December 2006, the Chinese river dolphin, also known as the baiji, or Yangtse river dolphin, was declared functionally extinct. A handful of individual baiji may still be alive, but not enough for species recovery. Dolphins are intentionally killed, legally and illegally, in many parts of the world and are unintentionally killed by entanglement in fishing lines and nets. Other formidable threats to dolphins worldwide include prey reduction because of overfishing and bycatch, climate change, and habitat degradation in the form of vessel harassment and collision, incidental mortality during fishing operations, oil spills, and anthropogenic noise emissions (from boats, oil exploration, and military activities). But by far the most widespread and seemingly detrimental force against dolphins is pollution. Studies of behavior and communication among dolphins are critical to understanding how these threats can best be mitigated.

The effects of global warming and climate change are complex and varied. Changes in ocean temperature and salinity will influence how sounds propagate through water. It is not a leap to conclude that this will affect dolphins' ability to communicate acoustically. Accordingly, dolphins—indeed, all marine animals that rely on vocal communication—may be required to modify the duration, frequency, or quality of their vocal signaling as aspects of their environment change. Whether this is possible is a question for the future.

Invisible synthetic underwater noise is perhaps the greatest impediment to communication for cetaceans. The issue of human-generated marine noise is becoming increasingly controversial as levels of this noise continue to grow in the world's oceans.[5] Activities such as military active sonar, shipping noise, seismic surveys, and Acoustic Thermometry of Ocean Climate (commonly referred to as ATOC) on whales and dolphins (and other marine life) have the potential to cause hearing loss as well as other physiological impacts. Sound is a pressure wave and can injure tissues throughout the body of an animal underwater. Various behavioral responses can include negative effects on the abilities of dolphins and other marine species to communicate with one another, let alone survive.[6]

Chronic and persistent ambient noise in the dolphins' world is often related to increases in oceanic vessel traffic. Some research suggests that shipping noise has impeded the range through which important cetacean vocalizations related to feeding and communication can be heard and perhaps to group dynamics as a whole.[7] In the busy coastal waters off California, noise from shipping has increased roughly three decibels per decade and is increasing even faster in other areas.[8] Shipping noise and seismic surveys have a high potential to interfere with cetacean communication via masking, which occurs when background noise overwhelms the signal of interest (such as a fellow whale's vocalization).[9]

Military active sonar, whether using high-, mid-, or low-frequency sound, generates a "ping" with a very loud source level. Because of the sporadic, short-term use of sonar during military exercises and combat situations, sonar pings are more likely to have acute rather than chronic effects on cetaceans. In addition, most active sonar is used in the mid-frequency range, which does not travel as far as low-frequency sound. Low-frequency active sonar is more likely to affect the communication of baleen whales, whereas mid-frequency active sonar has a particularly dangerous, and sometimes fatal, effect on deep-diving toothed cetaceans. A widely published series of mass strandings, particularly of beaked whales and other odontocetes, has led to increased scrutiny of the widespread use of mid-frequency sonar.[10]

Fortunately, countries around the world are increasingly designating marine protected areas to protect ecosystems; in some cases these areas are targeted toward cetaceans. Researcher Erich Hoyt has documented more than 350 marine protected areas. He comments, however, that "few of these areas are set up to protect dolphin communication, but they should be."[11] Although some areas effectively protect dolphins and whales from such loud noises as naval mid-frequency sonar, seismic exploration, and the increasing clamor of world shipping, Hoyt notes that "none of the areas will protect cetaceans from Navy *low*-frequency sonar, which can penetrate hundreds or even thousands of miles (kilometers) across the world ocean."[12]

Playing back sounds and monitoring the dolphins' responses can help to determine individual- and population-level impacts of anthropogenic noises.[13] Applications in wildlife management and conservation include mitigating interference with human industrial activities such as fishing; minimizing harm from other human activities such as seismic surveys and air-gun testing, drilling noise, and underwater explosions; and investigating avoidance responses of marine mammals to lessen the risk of vessel collisions.

Volker Deecke has reviewed the fascinating history of playback experiments with cetaceans and other marine mammals.[14] Experimental playbacks of calls to toothed cetaceans that were also applied to management situations began in the 1970s. Killer whale calls were used to exclude seals and sea lions from ongoing fishing areas, and studies involving the playbacks of conspecific calls to belugas in captivity and the wild were conducted. The 1980s brought an awareness of the need for, as well as the technology to, examine the effect of noise associated with offshore oil and gas extraction (air-gun sounds and drilling noise) on baleen whales. Since then, researchers have examined the effects of sound on dolphins from vessels as small as Jet Skis to large cruise and industrial, commercial ships.

The use of some of these technologies is not without various shades of controversy. One study concluded that killer whales were negatively affected and displaced in two areas after the introduction of acoustic deterrent devices into their habitat. While the devices were operating, both mammal-eating and fish-eating orcas declined in numbers. They returned to baseline levels when the devices were off. Invasive techniques such as playback experiments, acoustic deterrent devices, and tagging/bolting telemetry devices onto cetaceans have often helped protect these animals. However, as innovative and informative as they may be and, regardless of researchers' good intentions, sometimes the animals suffer. Scientists continue to search for more humane and benign methods of gathering information to study and respond to the conservation needs of whales and dolphins.[15] In June 2007, a journalist from Canada reported sighting an injured, undernourished beluga with a satellite transmitter clearly dragging through its skin, prompting a wildlife management board to consider changing how research is conducted in the Arctic so that no more animals are subjected to such "horrifying treatment."[16]

Dolphin and whale watching in the wild is a rapidly growing segment of the tourism market. In Shark Bay, Western Australia, a group of bottlenose dolphins allows close contact with waders who have regularly fed them for approximately forty years.[17] This group of dolphins and the associated feeding program has evolved into an important source of local economic revenue and has inspired dolphin-feeding enterprises elsewhere.[18] Other dolphin groups are targeted by large numbers of people seeking close contact without feeding programs. In New Zealand, for instance, tens of thousands of people attempt to swim with bottlenose and dusky dolphins from commercial boats every year.[19] Atlantic spotted and bottlenose dolphins have interacted with swimmers and divers in both near and offshore Bahamian waters fairly consistently for decades.[20] In Hawaii swimmers and kayakers regularly seek out spinner dolphins, even though swimming with dolphins in U.S. waters is illegal. In Japan people flock to swim with wild bottlenose dolphins that, ironically, are sometimes targeted by fishermen. Mikura Island saw its dolphin swim program grow exponentially from 1994 to 2000, when almost ten thousand people visited Miyake Island, 10 miles (18 km) north of Mikura, between May and September to ride boats that took them to swim with the dolphins around Mikura. I (Kathleen) began studying these dolphins in 1995. In 1998 and 1999, I saw the dolphins

In the United States, it is illegal to feed wild dolphins anything. All marine mammals, including dolphins, are protected by the Marine Mammal Protection Act of 1972.

become more agitated and somewhat aggressive toward people in the water. In 2000, Miyake's volcano erupted, causing a five-year evacuation of the island and a subsequent 90 percent reduction in the number of tourists visiting either island to swim with dolphins around Mikura. Over the next two years, the dolphins' behavior "relaxed" both toward swimmers and more generally when we were observing them. The Mikura fishing and tourism cooperatives have since adopted strict regulations regarding swimmer and boat interactions with the dolphins, benefiting both communities. Swim situations like those at Mikura often begin noncommercially, involving only a few people and dolphins, and subsequently become the focus of dolphin-swim businesses that can attract large numbers of tourists who want up-close meetings with dolphins.

Free-ranging, "friendly" dolphins who interact with people are frequently harassed, hurt, or even killed—whether intentionally or unintentionally. From the Caribbean to South America and Hawaii to Canada, I (Toni) have found that most recreational and professional boat operators earnestly believe they do not harass the dolphins they are watching. Yet I have seen many of them, especially in countries in which cetaceans are not legally protected, operate their boats in ways that are highly dangerous for the dolphins, such as pursuing them at high speeds, occasionally even separating mother-calf pairs. When a boat chases dolphins or when swimmers interrupt a group of resting or feeding dolphins, the dolphins may exhibit signs of disturbance, such as repetitive fluke slapping at the surface. A consumer-driven industry (with notable exceptions) combined with an overzealous desire to be near dolphins can certainly contribute to errors in judgment and sensitivity. When dolphin watching is conducted carefully and from a respectable distance, however, observable impacts to dolphins targeted for watching can be minimized.

Severe and lethal injuries from boats, cumulative and long-term changes related to breeding behavior, disruption of rest, displacement from habitat, and long-term responses to human activity (such as sensitization and habituation) have been associated with vessel and swimmer contacts.[21] In particular, increased mortality has been documented for solitary, sociable dolphins and other free-ranging dolphins who were regularly fed by humans. Researcher Lars Bejder has observed that in some areas, the impact of numerous swimmers was minimal, whereas in other habitats the impact of even one swimmer was enough to disrupt rest, feeding, and other important dolphin activities.[22] Similarly, Rochelle Constantine has found that over time the bottlenose dolphins in New Zealand's Bay of Islands are interacting with swimmers less and avoiding them more. David Lusseau has found that an entire population of dolphins is being threatened by the numerous tourist boats that come to see the dolphins in Milford Sound, also in New Zealand.[23] He has documented dolphins being injured and killed by the boats and states that up to 7 percent of them bear visible scars from collisions. Even large toothed cetaceans like sperm whales, another species vital to New Zealand's whale- and dolphin-watching industry, are sensitive to harassment from boat traffic.[24] The presence of whale-watching boats can cause whales to change their direction of swimming, likely causing unnecessary stress. Anna Forrest and S. Courbis have observed that Hawaiian spinner dolphins in Kealakekua Bay, Hawaii, may be avoiding important areas for resting, nursing, and mating because of the increased presence of swimmers and boaters.[25] Intensive whale-watching activity from boats appears to disrupt important activities like resting, feeding, and mating in seriously depleted populations of killer whales in North American waters.

Dolphins will often leap toward a moving vessel to catch a ride on the bow wave. This can be a responsible way of viewing dolphins in the wild.

The co-operator of Wild Side Tours and director of the Wild Dolphin Foundation on Oahu, Tori Cullins, has said that her business would be willing to sacrifice a drop in clientele rather than allow the number of human-to-dolphin contacts to continue climbing.[26] Not all tour operators share this sentiment, however. If the trend continues, the irony would be that we could love nature (in this case the dolphins) to death.

Even when dolphins do not show recognizable signs of stress in response to human contact, excessive proximity to people places them at higher risk of being deliberately or accidentally harmed. We must remember that the oceans, seas, and bays that we enter to visit dolphins are their homes as well as their playground. As we learn more about their communication systems and social lives, we will be better informed to manage our visits more appropriately. Once we know the proper etiquette, it will be our responsibility to be good guests. Until then, we should err on the side of caution.

Different cultures relate to dolphins in starkly contrasting ways today. Some view dolphins as little more than food, fertilizer, or a form of commerce, whereas others demonstrate a high degree of respect for these animals. For dolphin tourism to be humane, sustainable, and environmentally responsible, proactive measures must be taken to protect dolphins from negative impacts of human activities. To this end, a precautionary approach should be taken in legislation and management of watching activities, and regulations based on the best available scientific information should be developed and regularly monitored by independent and trained observers.[27] The tremendous socioeconomic potential for viewing free-ranging dolphins and whales in many areas of the world that rely on tourism as an industry has yet to be realized. Rather than irreversibly depleting natural resources, responsible watching has the potential to offer many economies an unprecedented opportunity to respond to a burgeoning number of environmentally sensitive naturalists and tourists while enhancing local stewardship of their natural environment.

Caution is warranted here because even well-intentioned attempts to view dolphins can affect individual animals and populations. Interruption of resting, feeding, nursing, and mating behaviors can occur easily, even with experienced tour operators. This is particularly true when guests have high expectations regarding their wildlife experience.[28] Over the past decade, there has been considerable research on the short-term effects of tourism industries on wild toothed whales, but studies designed to monitor long-term effects are rare. Short-term impacts are more readily measured and correlated to human activities than the more critical long-term effects.[29] Many studies have observed short-term reactions to human activities including changes in respiration and surfacing, swimming speed and course, group activity and social behavior, vocalizations, and individual behaviors suggestive of stress responses.[30]

A literature review that I (Toni) conducted for the International Whaling Commission revealed that odontocetes that have the highest degree of contact with humans in the wild are generally at the greatest risk of injury, illness, and death.[31] In particular, incidents in which humans intentionally injured or killed sociable odontocetes were reported almost exclusively for solitary dolphins as well as for those regularly fed by humans.[32] From Opo, the solitary dolphin who played with beachgoers and gave children "rides" but was later killed by a fisherman, to Luna, the friendly young orca who was separated from his pod and eventually killed in the propeller of a tugboat, life with humans can be a challenge. Although some cetaceans seem to enjoy human interaction, the long-term implications are that solitary, sociable individuals will be injured or killed—intentionally or accidentally—by humans and human activity.

Research that we both conducted with our colleagues Barbara Bilgre, Nicole Crane, and Alison Sanders on a bottlenose dolphin in Belize, as well as work that Toni did with Cathy Kinsman on belugas, has shown that successful protection of solitary odontocetes and humans with whom they associate is directly related to early implementation and consistency of on-site, proactive stewardship, management, and research programs.[33] Clearly, research and management are warranted to protect both the odontocetes and humans who participate in these astonishing, though potentially dangerous, situations.[34]

Observations from studies on captive swim-with-the-dolphin programs indicate that these activities may not always have the best effects on participating dolphins. Three of the four published studies that focus on behavioral indicators of stress in captive dolphin swim programs all report stress-related and avoidance-related behaviors from dolphins in the presence of swimmers. In the early 1990s, I (Toni) observed captive dolphins direct behaviors toward swimmers that were related to stress and aggression.[35] Within a year or so, the federal government commissioned a second, broader study to examine swim-with-the-dolphin programs at four American facilities. With respect to high-risk behaviors, the investigators found that captive dolphins frequently behaved submissively to swimmers even when the swimmers were small in stature, minimally mobile, and not aggressive.[36] The third study, on captive swim programs in New Zealand, showed that dolphins spent significantly more time in a swimmer-free refuge area during swim programs than when there were no swimmers in the pool, suggesting that the dolphins were actively avoiding swimmers.[37] The most recent study, also in the United States, is the only one that did not reveal notable behaviors indicative of stress related to swim programs.[38] The focus of this study differed slightly from the other three in that it focused primarily on behaviors indicative of positive welfare—specifically,

play-related behaviors. Perhaps the variation in results from studies on these programs elucidates the differences that can be observed between different animals
and environments.

The scant research on petting and feeding programs is a loud cry for more
rigorous examination into how these programs affect dolphin welfare.[39] Signs of
poor welfare vary even among experts, but the available data suggest that numerous interactions were directly harmful to both dolphins and humans; for example,
dolphins bit members of the public, and people fed dolphins inappropriate foods.
Even though these data are limited, they indicate the importance of proceeding
with extreme caution when considering a captive feeding program.

Many countries have few, if any, legal requirements governing the welfare and
unique needs of captive dolphins or of the protocol to be followed within swim
programs, either in the wild or in captivity. This often results in diminished welfare and even death for the animals.[40] In the United States, the Marine Mammal
Protection Act of 1972 governs all human interactions with wild dolphins, whales,
and porpoises, as well as seals, sea lions, sea otters, polar bears, and manatees. It is
illegal to change the behavior of, harass, harm, or kill any marine mammal. Because
swimming with a marine mammal typically results in a change of behavior in that
animal, it is thus not legal in U.S. waters. Other countries, such as Chile, prohibit
the capture or confinement of these cetaceans, whereas nations such as Italy and
Brazil prohibit interactive programs involving physical contact and public feeding.
Increasing numbers of countries, such as the United Kingdom and Australia, are
phasing out captive public display facilities, while other nations have denied permits
to capture, import, or export dolphins, Mexico being the most recent example.

Some countries, such as Japan, still allow dolphin hunting. Working against this
practice is Act for Dolphins (AFD; www.actfordolphins.com), a group of academics, scientists, and marine mammal professionals that opposes the unnecessary,
unsustainable, and inhumane treatment of dolphins and other small cetaceans killed
annually during drive hunts. These hunts are regulated by the Japanese government
and conducted by groups of fishermen who herd hundreds, sometimes thousands,
of dolphins and other small cetaceans into shallow bays. The dolphins are then cor-

ralled into nets and dispatched in a brutal manner. The methods, say researchers, result in a slow, painful death for these intelligent marine mammals.

Cofounder Lori Marino notes, "We hope [AFD] serves as a model for more involvement of the scientific community (and students) in welfare issues."[41] According to the organization, "the ethical argument for ending the drive is supported by a solid foundation of scientific evidence indicating that dolphins possess the mental and emotional capacities for pain and suffering on par with great apes and humans. It is also increasingly clear that dolphins have social traditions and cultures, complex interdependent relationships, and strong family ties all of which are susceptible to disruption or even dissolution in the drives." Many of the species hunted in Japan are on the World Conservation Union's Red List of Threatened Species. The hunts have also drawn criticism from relevant management organizations on both conservation and welfare grounds, including the International Whaling Commission, the treaty organization that regulates the hunting of the great whales.

The public display of captive dolphins, particularly those used in interactive programs, has become increasingly controversial in both scientific and public arenas.[42] Population and conservation biologists, as well as government agencies internationally, are concerned about potential effects from the increasing numbers of facilities in which dolphins are held captive. Dolphins have been kept in captivity since the first century C.E., when a stranded killer whale was captured and kept by Roman guards for sport.[43] Private collections of dolphins held in tubs and pools were attempted as early as the 1400s. In the mid-1800s, P. T. Barnum publicly exhibited captive bottlenose dolphins and belugas in a New York museum. Facilities dedicated to this purpose began to burgeon in the late 1930s. It is estimated today that there are at least two hundred captive dolphin exhibits in sixty or more countries.[44] The smaller toothed whales and dolphins are most commonly kept in human care; bottlenose dolphins are the most adaptable and readily habituated odontocete to the captive environment. Still, some facilities include killer whales, belugas, false killer whales, Pacific white-sided dolphins, and others.

Captive marine mammals have traditionally been simply on display or used to perform trained behaviors for the public. Over the past several decades, the popularity of interactive programs that allow tourists to swim with, wade with, touch, and feed captive dolphins has created a tremendous increase in the international market for capture of these animals from the wild. Swim-with-the-dolphin programs proliferate in part because of large tourist revenues generated from the cruise ship industry, which brings tourists to interact with captive dolphins as part of their recreational itineraries. Swim-with-the-dolphin facilities catering to tourists are opening in the Caribbean at the rate of two new programs per year.

One benefit these programs can offer is to expand their educational initiatives to provide the most accurate and up-to-date information about dolphin social lives and communication; several companies already offer strong educational programs that present opportunities for controlled interactions between dolphins and human visitors. Programs like these should lead to increased public awareness of marine mammal conservation and protection issues.

When we consider animal welfare, we refer to the degree of well-being or suffering that the animal experiences.[45] *Well-being* refers to a positive state, whereas *suffering* refers to a prolonged or acutely negative state. When an animal is subjected to aversive stimuli or deprived of certain behavioral opportunities, its well-being is threatened and suffering may result.[46] A thorough appraisal of an animal's welfare considers its mental, physical, and physiological condition. This assessment is optimally conducted by combining behavioral data with physical and physiological data; these data may be useful only when observed in conjunction with behavioral parameters.[47] Behavioral research on dolphins during interactive programs, therefore, is important in evaluating the welfare of dolphin participants and thus for minimizing risks to dolphins and humans. Behavioral assessment of dolphin welfare is also important for the husbandry and management of captive dolphins.[48]

Behavior is often one of the only available indicators of the condition of marine mammals. Recognizing and evaluating behavioral changes in dolphins is critical to identifying pre-pathological states.[49] Measurements of physiological stress may be difficult or undesirable to obtain because of concerns about increasing an animal's stress, especially in aquatic animals. It is also important to remember that the nature of the stressor does not always correspond to the manifestation of stress.[50] For example, physical abnormalities such as self-inflicted injuries may result from an animal's response to psychological stressors. The severity of the stress response, too, is not determined by the nature of the stressor. For example, psychological stimuli can produce an equal or greater stress response in animals than physical stressors.[51]

Abnormal behavior in an animal may be a product of stressful circumstances and may indicate suffering.[52] By abnormal behaviors we mean actions that a minority of the population performs.[53] Yet the absence of abnormal or other potentially stress-related behaviors should never be used to conclude that an animal is not suffering physically or psychologically. It is not uncommon for captive dolphins to display behaviors interpreted as abnormal, including unusually stereotyped behavior, self-destructive behavior, self-mutilation, and excessive aggression toward humans and peers.[54]

At least two assessments compare what we know about viewing dolphins in the wild and in captivity.[55] Although different in scope, both conclude that viewing cetaceans in the wild, when conducted in a responsible and precautionary manner, offers more benefits to and fewer negative effects on both dolphins and people, and can provide a uniquely important form of tourism and income to local communities. (A side note: Kathleen is currently conducting a comprehensive study of wild and captive dolphin behavior employing identical methods for data collection and analyses. Preliminary findings suggest more similarity than difference between captive and wild dolphins when examined at the individual level.) Watching cetaceans in the wild offers an unparalleled combination of educational, socioeconomic, and conservation benefits as well as an opportunity for environmentally sensitive nature-based tourism. Still, in my (Kathleen's) opinion, the value of being able to interact with dolphins, even in a captive swim-with-the-dolphin program, cannot be dismissed outright; I have witnessed how these interactions succeed in fostering people's greater awareness of conservation issues.

The value and impact of viewing dolphins in the wild as well as in captivity has yet to be adequately and scientifically examined. Consequently, many researchers have concluded that a precautionary principle should be applied to dolphin-human interactions. Perhaps the most valuable aspect of watching cetaceans in the wild is the potential to educate people to value marine mammals and their environment by providing a first-hand experience of these animals in the habitat in which they live. Alas, relatively few people (compared with the total human population) will have the opportunity to visit dolphins in the wild; therefore, captive dolphins, when well cared for in an enriched environment, can act as ambassadors for their wild counterparts. Despite the need for data documenting educational benefits, existing data do suggest that whale watching in the wild can "foster more appreciative and concerned attitudes toward whales."[56] When conducted responsibly and educationally, dolphin- and whale-watching tours teach people about the importance of maintaining the habitat of these animals and inspire their greater involvement in conservation efforts.[57] Dolphins and whales are charismatic megafauna, the darlings of the media and many people. If we protect them, we often protect many other marine species that might be less charismatic.

Kathleen has written elsewhere: "It is important to realize that the ocean is not our home, but our playground."[58] Not only do we both consider this a poignant and eloquent declaration reflective of her many years spent underwater with dolphins, but it mirrors the bumper sticker on Toni's car, which reads, "The ocean is my playground." We couldn't agree more. "We are guests and should act accordingly and with caution. Even though equipped with a huge smile and a seemingly forever-harmonious disposition, dolphins and whales are wild animals and should be treated with respect. If our future with dolphins is to include real interspecies communication, respecting dolphins enough to 'listen' to them would be the first step."[59]

The better we understand the signals used to coordinate communication and social activity among individuals within a population, whether dolphins or elephants or chimpanzees, the better we will understand the evolution of that population's social life and strategies. Increasing our knowledge of dolphin social ecology and communication heightens our understanding of the dynamics of their society and how we—human interlopers into their underwater world—affect their development as individuals and as a community. We can hope that greater understanding of what lies behind the dolphin's smile will lead to better conservation and protection for these amazing ocean-dwelling animals.

Appendix I A Behavioral Guide to Dolphin Signals

These behaviors represent a brief dictionary or vocabulary that defines what we understand about various dolphin postures, gestures, and behaviors. Some of these actions are depicted in sketches or photos in the text.

One needs to view the belly area of the dolphin to identify males from females.

female — umbilicus, mammary slit, vaginal opening, anus

male — umbilicus, opening for penis, anus

The belly of dolphins provides a view to distinguishing between females and males. The males have a sexually dimorphic space between the anus and genital slits. The females have mammary slits parallel to the urogenital slit.

probe — towards tail = male

POSITIONS OF INDIVIDUALS OR GROUPS

Single Dolphin

Hang	Stationary with the tail lower than the head (at an approximately 45-degree angle) just under the surface
Horizontal	Head and fluke are positioned in the same horizontal plane
Lying on bottom	Lying flat on the sea floor
Onside	Body is turned to one side (eye to surface) when swimming
Rest	Stationary at surface in horizontal position
Sink	Slowly moves deeper into the water column, any position
Somersault	Forward flip underwater
Spyhop	Head comes up and out of water to pectoral fins before retracing path and reentering the water

Two or More Dolphins

Above	Above a peer in the same vertical plane
Ahead	Ahead of a peer, in the same orientation but to the left or right of the other dolphin's rostrum
Contact	Body-body physical contact (dolphin positions identified)
Contact position	Pectoral fin of one dolphin is placed on the side between dorsal fin and tail of another dolphin with no fin movement
Head-to-head	Dolphins are head-to-head with respect to each other
Parallel	Dolphins are next to each other with same body orientation

BEHAVIOR CODES

Actions of Individuals

Bubble stream	Bubbles are produced in a stream from the blowhole
Fluke slap	Flukes are slapped at the water's surface

In The Bahamas, Atlantic spotted dolphins rub into the sand for various reasons that include scratching an itch, removing tiny parasites, or maybe because it just feels good.

Head nod	Head and rostrum move abruptly vertically or horizontally
Head scanning	Head moves side to side (often seen during echolocation)
Jaw clap	Jaws open and close abruptly and forcefully
Sand rubbing	All or part of the body is rubbed in the sand

Interactions between Two or More Dolphins

Approach	One dolphin approaches another
Belly-to-belly	Two dolphins swim belly to belly
Bite	One dolphin bites or rakes its teeth on another dolphin
Chase	One or more dolphins swiftly follow other dolphin(s)
Circle chase	Dolphins circle each other while swimming fast
Petting	One dolphin moves its pectoral fin along another dolphin's pectoral fin
Rubbing	One dolphin rubs its body on another dolphin's body
Synch petting	Two dolphins pet a third dolphin simultaneously
Synch breathe	Two or more dolphins synchronously surface and breathe
Fluke hit	One dolphin hits another dolphin with its fluke

Appendix 2 Etiquette for Interacting with Dolphins

Appropriate human conduct in the presence of dolphins differs widely owing to the varying needs of different dolphin species, social groups, and environments. Accordingly, government authorities around the world often enact laws or make recommendations that may reflect these and other factors. The suggested etiquette provided here is not intended to supersede any set rules or regulations. However, the safest (and most polite) way to encounter dolphins is to take the precautionary approach—just as we might when visiting someone's home whose culture is very different from our own.

The following code of conduct is a general guide for appropriate human behavior when interacting with dolphins. In providing this list, we are not encouraging close contact with dolphins, and even following these guidelines will not ensure the safety of humans or dolphins. These recommendations were written for programs offering swims with wild dolphins but also apply to programs where people can interact with dolphins in captivity. In both cases, it is important to listen to the educators and trainers leading the program.

Do not try to touch the dolphins. If the dolphins want physical contact with people, they will initiate it. If you try to touch one dolphin, often all the dolphins leave the area. Not only does this adversely affect dolphin behavior, but it irritates the other people with your group since everyone loses on the encounter.

Do not chase or swim after the dolphins. There is no way any swimmer can keep up with a dolphin. Swimming after dolphins when they move off is not only rude but may disturb them and force them to move farther away.

Always listen to your dolphin guide, captain, trainer, or educator. Not only do these individuals often have many hours (or years) of experience around dolphins, but they also have extensive experience on and in the water. If they guide you to a certain direction or approach, swim that way. It will usually be the best place to see the dolphins.

Remember your limits. Don't push yourself—enjoy your experience. If you are not confident about your skin-diving abilities or don't feel like getting in the water, *don't do it.* You can have a lot of fun watching from the boat or from land and sometimes even see a more complete picture of what the dolphins are doing.

When in the water, look down. The dolphins often swim right underneath you or approach you from the side or from behind.

Be willing to snorkel below the surface, but be careful and aware of your limitations. If you feel uncomfortable in the water, ask your dolphin guide or trainer for suggestions. Snorkeling is a fun sport, but only when one is comfortable. This will also enhance your experience with the dolphins.

When surfacing to breathe, look up to avoid any potential objects in your way.

Never try to feed wild dolphins.

Listen underwater. Often you can hear the vocalizations of dolphins and have a wonderful time eavesdropping on dolphin conversations without even seeing them.

Be aware of the boat when you are in the water. Be ready to enter the water and be ready to exit when told to do so. When entering, go feet first, holding your mask against your face with your snorkel in your mouth. The exit from each boat is different, so be sure to listen to your captain and dolphin guide for directions.

If you are an experienced skin-diver, you may want to try one or more of the following tips:

- The best approaches underwater to swimming with dolphins seem to be from an angle that is from the side or at about 45 degrees (roughly). Direct or head-to-head approaches or fast swims at 90 degrees to others usually signify aggressive activity.

- Try dolphin kicking (like butterfly swimming with both legs together). It is fast but requires a lot of energy.
- Underwater turns, circle swimming, and flips may indicate play to dolphins. They may try to imitate you doing these behaviors.
- Use your body as a tool to communicate. Remember the streamlined shape of the dolphins.

Always give dolphins space. Do not approach dolphins closely, either in boats or when swimming. They will come to you if they wish to interact.

Leave the water immediately if any dolphin exhibits violent or unusual behavior.

Remember, if the dolphins seem interested in us, they will come to swim near us. We are visitors to their world and should not abuse their welcome. Let's reciprocate by not infringing on their activity and by not polluting the oceans. With these simple guidelines and a gentle, courteous attitude, we can show our respect for the ocean and all the creatures that live in it.

Appendix 3 Scientific Names of Species Mentioned in the Text

blue whale (*Balaenoptera musculus*)

sperm whale (*Physeter macrocephalus*)

beluga (*Delphinapterus leucas*)

Commerson's dolphin (*Cephalorhynchus commersonii*)

Hector's dolphin (*Cephalorhynchus hectori*)

Indo-Pacific hump-backed dolphin (*Sousa chinensis*)

tucuxi (*Sotalia fluviatilis*)

common bottlenose dolphin (*Tursiops truncatus*)

Indo-Pacific bottlenose dolphin (*Tursiops aduncus*)

pantropical spotted dolphin (*Stenella attenuata*)

Atlantic spotted dolphin (*Stenella frontalis*)

spinner dolphin (*Stenella longirostris*)

short-beaked common dolphin (*Delphinus delphis*)

long-beaked common dolphin (*Delphinus capensis*)

white-beaked dolphin (*Lagenorhynchus albirostris*)

Atlantic white-sided dolphin (*Lagenorhynchus acutus*)

Pacific white-sided dolphin (*Lagenorhynchus obliquidens*)

dusky dolphin (*Lagenorhynchus obscurus*)

killer whale (*Orcinus orca*)

long-finned pilot whale (*Globicephala melas*)

short-finned pilot whale (*Globicephala macrorhynchus*)

rough-toothed dolphin (*Steno bredanensis*)
Amazon river dolphin (*Inia geoffrensis*)
Chinese river dolphin, or baiji (*Lipotes vexillifer*)
harbor porpoise (*Phocoena phocoena*)
vaquita (*Phocoena sinus*)
Dall's porpoise (*Phocoenoides dalli*)

sea otter (*Enhydra lutris*)
chimpanzee (*Pan troglodytes*)
African elephant (*Loxodonta africana*)
hippopotamus (*Hippopotamus amphibius*)
yellowtail jack (*Seriola dorsalis*)
African lion (*Panthera leo*)
vervet monkey (*Chlorocebus pygerythus*)
great tit (*Parus major*)
gorilla (*Gorilla gorilla*)
bonobo (*Pan paniscus*)

Notes

Chapter 1. A Dolphin's Life

1. Genus and species names can be found in appendix 3.
2. Frohoff 1996.
3. Kathleen Dudzinski, personal data and observations, 1997–2007.
4. See Perrin et al. 2002 for more details.
5. Baleen is unique to mysticetes. It is composed of keratin (the same substance that makes hair, nails, and horns in many other animals), but in these whales, it occurs in plates of bristles in the mouth where teeth might be found otherwise. These bristles form a filter through which food (such as krill or small fish) can be retained in the mouth while water is expelled through the baleen.
6. Reynolds and Rommel 1999; Perrin, Würsig, and Thewissen 2002.
7. Reynolds, Wells, and Eide 2000.
8. Reiss and Marino 2001; Herman, Morrel-Samuels, and Pack 1990; Herman et al. 1999; Herman et al. 2001.
9. Reiss and Marino 2001.
10. Rendell and Whitehead 2001:309.
11. Pryor 1973.
12. Untapped ecological niche: Barnes 1984.
13. Berta and Sumich 2004.
14. Thewissen and Williams 2002; Thewissen et al. 2006.
15. Thewissen et al. 2006.
16. Berta and Sumich 2004.
17. Thewissen et al. 2006.
18. Nikaido, Rooney, and Okada 1999.

19. Barklow 1997, 2004.

20. Barklow 1994, 1995, 2004.

21. Barklow 2004.

22. Barklow 1997, 2004.

23. Hof et al. 2000.

24. Langbauer et al. 1989.

25. Marino 2004.

26. Marino 2004; Hof, Chanis, and Marino 2005.

27. Marino 2002, 2004.

28. Social ecology: Connor et al. 1998; communication: Jerison 1973.

29. Mirror self-recognition: Reiss and Marino 2001; Marten and Psarakos 1995; abstract concepts: Herman, Richards, and Wolz 1984; Herman et al. 2001; learning: Janik and Slater 2000; culture: Rendell and Whitehead 2001.

30. Watkins and Wartzok 1985.

31. Mass and Ya 1995; van der Pol, Worst, and Andel 1995; Mass and Supin 2002.

32. Tarpley and Ridgway 1994.

33. Herman et al. 1975; Norris et al. 1994.

34. Herman et al. 1975.

35. Mass and Supin 2002.

36. Watkins and Wartzok 1985.

37. Van der Pol, Worst, and Andel 1995.

38. Pryor 1990a; Herzing 1990.

39. Würsig, Kieckhefer, and Jefferson 1990.

40. Perrin 1970; Herzing 1991a.

41. Leatherwood and Reeves 1983.

42. Würsig and Würsig 1980.

43. Norris et al. 1994.

44. Geraci 1986.

45. Pryor 1990a.

46. Worthy and Edwards 1990.

47. Perrin, Würsig, and Thewissen 2002.

48. Ridgway, Scronce, and Kanwisher 1969.

49. Polasek and Davis 2001.

50. Williams et al. 1992.

51. Reynolds and Rommel 1999; Thewissen, Williams, and Hussain 2000.

52. Caldwell and Caldwell 1977.

53. Clark and Mangel 1986.

54. Gubbins 2002.

55. Sargeant et al. 2005.

56. Personal communication between Alejandro Acevedo and Kathleen Dudzinski in mid-1990s and in 2006 to invigorate Kathleen's aging memory.

57. Dudzinski 1998; Dudzinski et al. (submitted).

58. Ballance 1990.

59. Bottlenose dolphin groups: Leatherwood and Reeves 1983; companions change: Wells 1991.

60. Wells, Scott, and Irvine 1987; Connor 1990a, 1990b.

61. Ford et al. 1999.

62. Shane 1990.

63. Aggressive behavior: Norris 1969; Pryor and Shallenberger 1991 with sexual behavior: Taylor and Saayman 1972; Pryor and Shallenberger 1991; Saayman and Taylor 1979.

64. Östman 1994.

65. Simard and Gowans 2004.

66. Bel'kovich 1991.

67. Bel'kovich 1991.

68. Dudzinski, unpublished data 1997–2006; Paulos, Dudzinski, and Kuczaj 2008.

69. Bel'kovich 1991.

70. Fagan 1981; Bekoff 1984.

71. Play variable: Bekoff 1972; environmental contexts: Bekoff 1995.

72. McBride and Hebb 1948; Morgan 1968; Taylor and Saayman 1972; Norris and Dohl 1980; Bel'kovich 1991; Hankins 1993.

73. Bel'kovich 1991.

74. Dudzinski, Douaze, and Thomas 2002.

75. Wells, Irvine, and Scott 1980; Johnson and Norris 1986.

76. Bottlenose dolphins: Shane 1977; Würsig and Würsig 1979; Wells, Scott, and Irvine 1987; Connor 1990a, 1990b; harbor porpoise: Gaskin 1982; short-finned pilot whales: Heimlich-Boran 1988.

77. Atlantic spotted dolphins: Herzing 1993; Dudzinski 1996; Hawaiian spinner dolphins: Norris et al. 1994; Östman 1994; dusky dolphins: Würsig and Würsig 1980; Cipriano 1992; humpback dolphins: Saayman and Taylor 1979; Karczmarski, Thornton, and Cockcroft 1997; killer whales: e.g., Ford et al. 1999.

78. Evans 1987; Grier and Burk 1992; Frohoff 2007.

Chapter 2. The Expressive Dolphin

1. This action is what animal behaviorists might consider an example of "displacement behavior."
2. For in-depth information about communication and its evolution, see Hauser 1996 and Bradbury and Vehrencamp 1998.
3. Consortship pops: Connor and Smolker 1996; maternal thunks: McCowan and Reiss 1995a.
4. Von Frisch 1967.
5. Poole 1996:143.
6. Dudzinski 1998; Herzing 1991b.
7. Dudzinski 1998.
8. Herman and Tavolga 1980; Caldwell, Caldwell, and Tyack 1990.
9. Unique to dolphins: Podos, Da Silva, and Rossi-Santos 2002; across species: Steiner 1981; Wang, Würsig, and Evans 1995a; Rendell et al. 1999; across geographic locations: Bazúa-Durán and Au 2004; across social groups: Janik 2000; across populations: Azevedo and Van Sluys 2005; Wang, Würsig, and Evans 1995b; Morisaka et al. 2005; different gender and age: Sayigh et al. 1995.
10. Caldwell and Caldwell 1977; Herman and Tavolga 1980.
11. Au 1993.
12. Reidenberg and Laitman 1988.
13. Cranford, Amundin, and Norris 1992.
14. Dawson 1991.
15. Ketten 1992; Cranford, Amundin, and Norris 1992.
16. Hemila and Nummelaa 1999.
17. Hemila and Nummelaa 1999; Vater and Kössl 2004.
18. Hemila and Nummelaa 1999.
19. Thewissen 2002.
20. Au 1993.
21. Spinka, Newberry, and Bekoff 2001.
22. Group hunting: Würsig and Würsig 1980; crater feeding: Rossbach and Herzing 1997; mud banks: Petricig 1993; Gubbins 2002; beaches: Sargeant et al. 2005; stun fish: Wells, Scott, and Irvine 1987; Connor et al. 2000; corral fish: Fertl and Würsig 1995.
23. Guinet and Bouvier 1995; Baird and Whitehead 2000.
24. Petricig 1993, 1995; Gubbins 2002, see review by Silber and Fertl 1995.
25. Sargeant et al. 2005.
26. Fedorowicz, Beard, and Connor 2003.
27. Gazda et al. 2005.

28. Packer 1994.

29. Moehlman 1989.

30. For example, Connor and Smolker 1985; Wells, Scott, and Irvine 1987; Wells 1991; Fertl 1994; Rossbach and Herzing 1997.

31. Mann and Smuts 1999.

32. Hendry 2003.

33. Kuczaj and Highfill 2005; Kuczaj and Yeater 2006.

34. Mann et al. 2000.

35. Yurk et al. 2002.

36. Ford 1989, 1991.

37. Ford 1989, 1991.

38. Poole 1996.

39. Nousek et al. 2006.

40. Norway: Similiä 1997.

41. Bain 1986; Baird 2000.

42. Moss 1982.

43. Poole 1996.

44. Caldwell, Caldwell, and Tyack 1990; Lammers and Au 2003.

45. Caldwell and Caldwell 1965.

46. Caldwell and Caldwell 1968.

47. Dreher 1961; Dreher and Evans 1964.

48. Caldwell, Caldwell, and Tyack 1990.

49. McCowan and Reiss 1995b, 1995c.

50. Smolker, Mann, and Smuts 1993.

51. Sayigh et al. 1990.

52. Janik and Slater 1998.

53. Trivers 1971; Connor and Norris 1982.

54. Tyack and Sayigh 1997.

55. Janik and Slater 1998; Tyack and Sayigh 1997.

56. Trainer's whistle: Kathleen Dudzinski and trainers at Roatan Institute for Marine Sciences, Anthony's Key Resort, Roatan, Honduras, personal communication, 2003; truck noises: Poole 1996.

57. Moss 1982.

58. Wells, Scott, and Irvine 1987; Connor, Heithaus, and Barre 1999.

59. Seyfarth and Cheney 1999; Janik and Slater 1998.

60. Azevedo and van Sluys 2005.

Chapter 3. Eavesdropping

1. Goodall 1986.
2. Payne 1998.
3. Reynolds and Rommel 1999.
4. Kathleen Dudzinski's observations of spotted dolphins in The Bahamas, 1992–2002.
5. Moss 1982.
6. Dudzinski 1996.
7. Frohoff 1993, 1996.
8. Würsig and Würsig 1980.
9. Norris et al. 1994.
10. Frohoff 2004.
11. S-shaped posture: Norris et al. 1994.
12. Würsig and Würsig 1980; Fertl and Würsig 1995; Dudzinski 1996.
13. Connor and Smolker 1996.
14. Caldwell, Caldwell, and Tyack 1990; McCowan and Reiss 1995.
15. Norris and Dohl 1980.
16. Griffin 1986.
17. Dudzinski 1996; Herzing 2000.
18. Xitco and Roitblat 1996.
19. Gregg, Dudzinski, and Smith 2007.
20. Dawson 1991.
21. Norris et al. 1994.
22. Frohoff, Packard, and Benson 1996.
23. Dudzinski, Clark, and Würsig 1995.
24. Dudzinski and Newborough 1997.
25. Schotten et al. 2005.
26. Morton 1977.
27. Dudzinski 1998.
28. Waples and Gales 2002.
29. Dudzinski 1996, 1998.
30. Walther 1984.
31. Krebs and Dawkins 1984.
32. Walther 1984; Krebs and Dawkins 1984; W. J. Smith 1991.
33. Fagan 1981.
34. Norris et al. 1994; Goodall 1971; Patterson 1978; Fossey 1983.
35. Herzing 1991a.

36. Bernd Würsig, personal communication, 1998. Rochelle Constantine, personal communication, 1999.

37. Atlantic spotted dolphins: Herzing 1991b; Dudzinski 1996; bottlenose dolphins: e.g., Dudzinski, Frohoff, and Crane 1995; beluga whales: Frohoff, Vail, and Bossley 2005, 2006; captive dolphins: Defran and Pryor 1980; Herman 1980; Stan Kuczaj, personal communication, 2002.

38. Walther 1984; Krebs and Dawkins 1984; Pryor 1986; W. J. Smith 1991.

39. Pryor 1986:256.

40. Communication through play: Fagan 1981; play in development of communication: Bekoff 1972; Spinka, Newberry, and Bekoff 2001.

41. Beach 1945; Fagan 1981.

42. Norris and Dohl 1980; Herzing 1993; Frohoff 1996.

43. Sweeney 1990.

Chapter 4. Beyond the Dolphin's Smile

1. Biologist Raymond Coppinger in *Dogs That Changed the World*, "Part 1: The Rise of the Dog," first broadcast January 13, 2008, on PBS; directed and produced by Corinna Faith.

2. Skinner 1938.

3. Darwin 1965.

4. Lusseau and Newman 2004.

5. Herman 2006.

6. Connor and Mann 2006.

7. Fraser et al. 2006.

8. Marino 2002.

9. Marino, McShea, and Uhen 2004.

10. Manger 2006.

11. Marino 2004.

12. Marino 2004.

13. Ridgway 2000.

14. Marino et al. 2000.

15. Marino 2004.

16. Marino 2004.

17. Space needed for processing auditory information: Simmonds 2006; rejection of this idea: Pabst, Rommel, and McLellan 1999.

18. Manger 2006.

19. Lilly 1961, 1978.

20. Lilly 1967.

21. Lilly 1973.

22. Norris 1969; Pryor and Norris 1991.

23. Miller 2003.

24. Schusterman and Gisiner 1988.

25. Pinker 1995.

26. Lilly 1961.

27. Batteau and Markey 1966.

28. Gardner and Gardner 1969; Premack 1971.

29. Bonobos: Savage-Rumbaugh et al. 1985; gorillas: Patterson 1978; parrots: Pepperberg and Brezinsky 1991.

30. Herman, Richards, and Wolz 1984.

31. Herman 2006.

32. Herman, Richards, and Wolz 1984.

33. Controversy: Schusterman and Gisiner 1988.

34. Herman 2006.

35. Visual and echolocation systems: Harley, Roitblat, and Nachtigall 1996; episodic memory: Dere et al. 2006; event memory: Mercado et al. 1999; perceive and classify objects: Mercado et al. 1999.

36. Herman, Morrel-Samuels, and Pack 1990.

37. Pryor 1973, 1986; Herman 2006.

38. Connor, Smolker, and Bejder 2006.

39. Herman 2006.

40. Fellner, Bauer, and Harley 2006.

41. Herman et al. 1999.

42. Tschudin et al. 2001.

43. Chimpanzees: Leavens 2004; dogs: Miklósi et al. 1998.

44. Xitco, Gory, and Kuczaj 2001.

45. Xitco, Gory, and Kuczaj 2004.

46. Dudzinski et al. 2003.

47. Pack and Herman 2007.

48. Xitco and Roitblat 1996.

49. Pack and Herman 2006.

50. Consciousness explained: Dennett 1991; present only in humans: Wynne 2004; found in nonhuman animals: Allen 1998.

51. Herman et al. 2001.

52. Xitco 1988.

53. Kuczaj and Yeater 2006.

54. Reiss and Marino 2001.

55. Gallup 1970.

56. Elephants: Plotnik, de Waal, and Reiss 2006; great apes: Povinelli et al. 1997.

57. Dolphins distinguish: Marten and Psarakos 1995; self-aware: Hart and Whitlow 1995.

58. Povinelli and Giambrone 2001.

59. One trial: Tschudin 2001; wrong box: Tschudin 2006.

60. Rendell and Whitehead 2001.

61. Roper 1986.

62. Killer whales: Baird 2000; bottlenose dolphins: Connor, Heithaus, and Barre 2001.

63. Ford 1989.

64. Ford 1991.

65. Deecke, Ford, and Spong 2000.

66. Christal, Whitehead, and Lettevall 1998.

67. Weilgart and Whitehead 1993.

68. Rendell and Whitehead 2003.

69. Ford 1991.

70. Bain 1986.

71. Bowles, Young, and Asper 1988.

72. Guinet 1992.

73. Guinet and Bouvier 1995.

74. Caro and Hauser 1992.

75. Rendell and Whitehead 2001.

76. Shark Bay dolphins: Connor and Mann 2006; observed since 1984: Connor, Smolker, and Richards 1992.

77. Connor and Mann 2006.

78. Connor, Smolker, and Richards 1992.

79. Harcourt and de Waal 1992.

80. Kudo and Dunbar 2001.

81. Connor and Krützen 2003.

82. Connor and Mann 2006.

83. Herman 1980.

84. Alexander 1979.

85. Connor and Mann 2006.

86. Smolker et al. 1997.

87. Biro et al. 2003; Ottoni and Mannu 2003; Weir and Kacelnik 2006.
88. Krützen et al. 2005.
89. Smolker et al. 1997.
90. Pryor et al. 1990.
91. Rendell and Whitehead 2001.
92. Marino 2002.
93. Darwin 1936[1871]:38.
94. Marten et al. 1996; McCowan et al. 2000.
95. Kuczaj and Highfill 2005.
96. Hurley and Nudds 2006.
97. Bekoff 2000a, 2000b.
98. Lautin 2001.
99. MacLeon 1970.
100. Aggleton 1992.
101. Marino 2004.
102. Allen and Bekoff 1997.
103. Bekoff 1995.
104. Bekoff 2000a.
105. Bekoff 2000a.
106. Douglas-Hamilton et al. 2006.
107. Post-traumatic stress disorder: Bradshaw et al. 2005; traumatic situations: Bradshaw and Schore 2007.
108. Williams and Lusseau 2006.
109. Bekoff 2000b.

Chapter 5. Where Humans and Dolphins Meet

1. Twiss and Reeves 1999; Frohoff 2007.
2. For example, Lilly 1975.
3. With the editorial assistance of his sons, Philip Hansen Bailey and Charles Lilly, in Frohoff and Peterson 2003:84.
4. Nollman 1987; Doak 1981, 1988; Wilke, Bossley, and Doak 2005.
5. Beamish 1995.
6. Lilly 1975; Nollman 1987.
7. Frohoff 1993.
8. Frohoff and Packard 1995; Frohoff 1996.

9. Boat activity: Au and Perryman 1982; Acevedo 1991; Kruse 1991; tuna-fishing operations: Norris, Stuntz, and Rogers 1978; Pryor and Shallenberger 1991.

10. Acevedo 1991; Richardson et al. 1991.

11. Purton 1978; Martin and Bateson 1986.

12. For example, Constantine 2001; Bejder and Samuels 2003; Lusseau 2003.

13. Many levels: Krebs and Dawkins 1984; Walther 1984; W. J. Smith 1991; change in behavior: Philips and Austad 1990; W. J. Smith 1990.

14. Krebs and Dawkins 1984; Walther 1984; W. J. Smith 1991.

15. Success of communication: W. J. Smith 1990; signals easily read: Walther 1984.

16. Krebs and Dawkins 1984.

17. Kiszka 2007:101.

18. Walther 1984:367.

19. Hediger 1964:163.

20. Pryor 1973; Pryor 1990b.

21. Dudzinski 2003.

22. Krebs and Dawkins 1984.

23. Dudzinski, Clark, and Würsig 1995.

24. For example, Xitco 1988; Herzing 1993; Kuczaj and Yeater 2006.

25. Frohoff, Vail, and Bossley 2005, 2006.

26. For example, Defran and Pryor 1980; Herman 1980; Herzing 1993; Dudzinski, Frohoff, and Crane 1995.

27. Frohoff 1998.

28. Fagan 1981.

29. Bekoff 1972.

30. Frohoff 1996.

31. Dudzinski 1996; Frohoff, personal observation, 1985, 1992.

32. Beach 1945.

33. We note that in some countries, including the United States, it is illegal to intentionally swim with wild dolphins (see chapter 6 for more information).

34. Norris and Dohl 1980.

35. Hankins 1993.

36. Donaldson 1982; Bel'kovich 1991; Herzing 1993.

37. Provisioned with food: National Marine Fisheries Service 1990; contact with swimmers: Lockyer 1990.

38. Samuels and Spradlin 1995.

39. Pryor 1986.

40. Pryor 1986; Frohoff 1993.

41. Frohoff 2000a, 2000b.

42. Pryor 1986.

43. Philips and Austad 1990; W. J. Smith 1991.

44. Pryor 1986.

45. Defran and Pryor 1980.

46. Frohoff 1993.

47. Frohoff 2007.

48. Differences: Taylor and Saayman 1972; similarities: Smuts 1988.

49. Taylor and Saayman 1972.

50. Smuts 1988.

51. Teleki 1972; Norris and Dohl 1980; Norris and Schilt 1988.

52. Herman 1980, 1986.

53. Worsham and D'Amato 1973.

54. Pack 1994.

55. Dolphins: Caldwell and Caldwell 1977; Nachtigal 1986; Kuznetsov 1990; Pryor 1990a; Würsig, Kieckhefer, and Jefferson 1990; primates: Goodall 1986.

56. For example, Box 1984; Würsig, Kieckhefer, and Jefferson 1990.

57. Taylor and Saayman 1972; Peters 1980; Connor 1990a, 1990b; Pryor 1990a.

58. Bel'kovich 1991.

59. Play with inanimate objects: McBride and Hebb 1948; as mechanism for sexual learning: Morgan 1968; directed toward other species: Bel'kovich 1991.

60. Frohoff and Peterson 2003; Gales, Hindell, and Kirkwood 2003.

61. Gales, Hindell, and Kirkwood 2003.

62. Gales, Hindell, and Kirkwood 2003.

63. Domning 1991; Pryor et al. 1990.

64. Lilly and Montagu 1963.

65. Frohoff, Vail, and Bossley 2006; Lockyer 1990; Wilke, Bossley, and Doak 2005.

66. Dudzinski, Frohoff, and Crane 1995; Frohoff 1996; Frohoff, Packard, and Benson 1996.

67. Hoyt 2002.

68. Samuels, Bejder, and Heinrich 2000; Gales, Hindell, and Kirkwood 2003; Samuels et al. 2003; Frohoff 2007, in press.

69. Frohoff 1996.

70. Marino and Lilienfeld 1998, 2007.

71. Humphries 2003.

72. Brensing, Linke, and Todt 2003:100.

73. B. Smith 2003; Frohoff 2003:63.

74. Marino and Lilienfeld 2007.

75. Buck and Schroeder 1990; National Marine Fisheries Service 1990; Mazet, Hunt, and Ziccardi 2004.

76. Mazet, Hunt, and Ziccardi 2004.

77. Mazet, Hunt, and Ziccardi 2004.

78. For example, "Dolphins Said to Have Rescued Fishermen," Reuters, October 25, 1993.

79. For example, "Dolphins to the Rescue: Fellow Mammals Repel Shark, Save Surfer," *Surfer Magazine*, 1993; "Dolphins Save Man in Shark Attack," *Washington Times*, July 26, 1996; "Columbia Stowaways Say Dolphins Saved Their Lives," Reuters, February 18, 1999.

80. For example, *Miami Herald*, January 23, 2000, April 22, 2001.

81. *New Zealand Herald*, November 24, 2004.

82. Frohoff 2000a; Gales, Hindell, and Kirkwood 2003.

83. Patterson et al. 1998.

84. Shane, Wells, and Würsig 1986; Wells, Scott, and Irvine 1987.

85. Hediger 1955.

86. Sweeney 1990.

87. Hediger 1964; Pryor 1973.

88. Pryor 1973.

89. Santos 1997.

90. Rose, Farinato, and Sherwin 2006.

91. Mazet, Hunt, and Ziccardi 2004.

Chapter 6. Communicating Conservation

1. Bauer and Herman 1986.

2. Defran and Pryor 1980; Pryor and Kang 1980; Au and Perryman 1982; Constantine 1995; Samuels and Spradlin 1995.

3. Norris, Stuntz, and Rogers 1978; Pryor and Kang 1980; Au and Perryman 1982; Finley et al. 1990; Acevedo 1991; Kruse 1991; Constantine 1995.

4. Constantine 1995.

5. Marine Mammal Commission 2007.

6. Richardson et al. 1998.

7. Soto et al. 2006.

8. Tyack 2003; Southall 2005.

9. Nowacek et al. 2007.

10. Marine Mammal Commission 2007.

11. Hoyt 2005; Erich Hoyt to Toni Frohoff, personal communication, July 24, 2007.

12. Erich Hoyt to Toni Frohoff, personal communication, July 24, 2007.

13. Deecke 2006.

14. Deecke 2006.

15. Morton and Symonds 2002.

16. Bob Weber, "Suffering Narwhal Opens Hidden Debate on Arctic Research Ethics," CP Wire, June 2, 2007.

17. For example, Connor and Smolker 1985.

18. Frohoff and Peterson 2003; Gales, Hindell, and Kirkwood 2003.

19. Constantine 1995.

20. For example, Herzing 1993; Dudzinski 1996.

21. Frohoff 2000b; Gales, Hindell, and Kirkwood 2003.

22. Bejder and Samuels 2003.

23. Lusseau 2003.

24. Richter, Dawson, and Slooten 2006.

25. Forest 2001; Courbis 2004.

26. Lisa Huynh, "Dolphins Enduring More Harassment: Officials Want Tougher Viewing Rules," West Hawaii Today; available at http://wilddolphin.org/westoahu.htm; Tori Cullins to Toni Frohoff, personal communication, 2007.

27. Frohoff 2000b; Bejder and Samuels 2003.

28. Duffus and Dearden 1992.

29. Bejder and Samuels 2003; Frohoff 2004.

30. For example, Frohoff 2000b; Constantine 2001; Williams, Trites, and Bain 2002; Bejder and Samuels 2003; Lusseau 2003.

31. Frohoff 2000b.

32. Samuels, Bejder, and Heinrich 2000; Samuels et al. 2003.

33. Frohoff 2000b.

34. Wilke, Bossley, and Doak 2005.

35. Frohoff 1993; Frohoff and Packard 1995.

36. Samuels and Spradlin 1995.

37. Kyngdon, Minot, and Stafford 2003.

38. Trone, Kuczaj, and Solangi 2005.

39. Frohoff and Peterson 2003.

40. Frohoff and Peterson 2003; Couquiaud 2005.

41. Lori Marino to Toni Frohoff, personal communication, July 16, 2007.

42. For example, Kellert 1996, 1999; Frohoff 2007.

43. Twiss and Reeves 1999.

44. For example, Couquiaud 2005; Rose, Farinato, and Sherwin 2006.

45. Dawkins 1990.

46. Hughes and Duncan 1988.

47. Dawkins 1980; Morton and Griffiths 1985.

48. Senate Select Committee on Animal Welfare 1985; Sweeney 1990.

49. Sweeney 1990.

50. Dawkins 1980.

51. Warburton 1991.

52. Meyer-Holzapfel 1968; Dawkins 1980.

53. Meyer-Holzapfel 1968; Dawkins 1980; Grier and Burk 1992.

54. Norris and Prescott 1961; Greenwood 1977; Defran and Pryor 1980; Carter 1982; Sweeney 1990.

55. Frohoff and Packard 1995; Carlson and Frohoff in preparation.

56. Kellert 1999:16.

57. Hoyt 2002.

58. Dudzinski 2003:295.

59. Dudzinski 2003:295.

Glossary

acoustic—of or relating to sound

analogous—having a similar function but appearing independently

anthropomorphism—the practice of assigning human qualities to nonhuman animals

Archaeocetes—the order of extinct, ancient whales

Cetacea—the order of living whales, dolphins, and porpoises, including both toothed and baleen whales

cognition—knowledge acquired through such processes as reasoning, intuition, and perception

culture—variations in behavioral repertoire passed on socially and unrelated to environmental or genetic variation

dialect—vocalizations specific to a region or group of individuals

eavesdropping—to listen in unnoticed on a conversation or to observe another individual or group unnoticed

echoic eavesdropping—to listen in on the echolocation of another individual

echolocation—clicks produced at high frequency by dolphins to investigate objects a short distance away or to find food, also known as sonar

ethology—the study of animal behavior and actions

fission-fusion society—a social structure in which small groups merge to form larger groups for socializing, foraging, and traveling and separate into smaller groups after the activity

flukes—the tail of a dolphin, porpoise, or whale

gustatory—of or relating to the sense of taste

homologous—sharing the same origin but having a different function

hydrophone—a microphone used to record sounds in the water

juvenile—a young animal, in dolphins between the ages of three and seven
      years old

lek-mating—practice in which individuals gather at a specific location to mate

matrilineal—following the mother's line, or matriline

melon—the forehead of a dolphin

MVA (mobile video/acoustic system)—the recording system Kathleen designed
      and built to record dolphin behavior and stereo sounds simultaneously

Mysticeti—baleen whales

Odontoceti—toothed whales and dolphins

olfactory—of or relating to the sense of smell

pectoral—the dolphin front limb or flipper

pelagic—relating to the open ocean

pod—a group of genetically related individuals

pulsed call—amplitude-modulated sound produced by dolphins

subadult—a dolphin teenager, an individual between the ages of about seven
      and eleven years old that is not yet socially or sexually mature

sympatric—having an overlapping geographic range

tactile—touch

thermoregulation—the maintenance of body temperature by conserving or
      releasing heat

vocalization—a sound produced by animals to communicate

whistle—a frequency-modulated pure tone produced by many dolphins

Bibliography

Acevedo, A. 1991. Interactions between boats and bottlenose dolphins, *Tursiops truncatus*, in the entrance to Ensenada de La Paz, Mexico. *Aquatic Mammals* 17 (3): 120–24.

Aggleton, J. P., ed. 1992. *The Amygdala: Neurobiological Aspects of Emotion, Memory, and Mental Dysfunction*. New York: Wiley-Liss.

Alexander, R. 1979. *Darwinism and Human Affairs*. Seattle: University of Washington Press.

Allen, C. 1998. The discovery of animal consciousness: An optimistic assessment. *Journal of Agricultural and Environmental Ethics* 10: 217–25.

Allen, C., and M. Bekoff. 1997. *Species of Mind: The Philosophy and Biology of Cognitive Ethology*. Cambridge, MA: MIT Press.

Au, D., and W. Perryman. 1982. Movement and speed of dolphin schools responding to an approaching ship. *Fishery Bulletin* 80: 371–79.

Au, W. W. L. 1993. *The Sonar of Dolphins*. New York: Springer-Verlag.

Azevedo, A. F., and M. Van Sluys. 2005. Whistles of the tucuxi dolphin (*Sotalia fluviatilis*) in Brazil: Comparisons among populations. *Journal of the Acoustical Society of America* 117: 1456–64.

Bain, D. 1986. Acoustic behavior of *Orcinus:* Sequences, periodicity, behavioral correlates and an automated technique for call classification. In *Behavioral Biology of Killer Whales*, ed. B. C. Kirkevold and J. S. Lockard, 335–71. New York: Alan R. Liss.

Baird, M. 2000. The killer whale: Foraging specializations and group hunting. In *Cetacean Societies: Field Studies of Dolphins and Whales*, ed. J. Mann, R. Connor, P. Tyack, and H. Whitehead, 127–53. Chicago: University of Chicago Press.

Baird, R. W., and H. Whitehead. 2000. Social organization of mammal-eating killer whales: Group stability and dispersal patterns. *Canadian Journal of Zoology* 78: 2096–105.

Ballance, L. T. 1990. Residence patterns, group organization, and surfacing associations of bottlenose dolphins in Kino Bay, Gulf of California, Mexico. In *The Bottlenose Dolphin*, ed. S. Leatherwood and R. R. Reeves, 267–83. San Diego: Academic.

Barklow, W. E. 1994. Big talkers. *Wildlife Conservation* 97 (1): 20–29, 80.

———. 1995. Hippo talk. *Natural History* 104: 54.

———. 1997. Some underwater sounds of the hippopotamus (*Hippopotamus amphibius*). *Marine and Freshwater Behavior and Physiology* 29: 237–49.

———. 2004. Amphibious communication with sound in hippos, *Hippopotamus amphibius*. *Animal Behaviour* 68: 1125–32.

Barnes, L. G. 1984. Search for the first whale: Retracing the ancestry of cetaceans. *Oceans* 17 (2): 20–23.

Batteau, D. W., and P. R. Markey. 1966. Man/dolphin communication final report. Prepared for United States Naval Ordnance Test Station, China Lake, CA.

Bauer, G. B., and L. M. Herman. 1986. Effects of vessel traffic on the behavior of humpback whales in Hawaii. Report to National Marine Fisheries Service, Honolulu, HI.

Bazúa-Durán, C., and W. W. L. Au. 2004. Geographic variations in the whistles of spinner dolphins (*Stenella longirostris*) of the main Hawai'ian Islands. *Journal of the Acoustical Society of America* 116: 3757–69.

Beach, F. A. 1945. Current concepts of play in animals. *American Naturalist* 79: 523–41.

Beamish, P. 1995. A "traffic light" for cetacean stress. In Abstracts, Eleventh Biennial Conference on the Biology of Marine Mammals, Orlando, FL, December 3–7, 1995, p. 9.

Bejder, L., and A. Samuels. 2003. Evaluating the effects of nature-based tourism on cetaceans. In *Marine Mammals: Fisheries, Tourism and Management Issues*, ed. N. Gales, M. Hindell, and R. Kirkwood, chap. 12. Collingwood, Australia: CISRO.

Bekoff, M. 1972. The development of social interaction, play, and meta-communication in mammals: An ethological perspective. *Quarterly Review of Biology* 47: 412–34.

———. 1984. Social play behavior. *BioScience* 34 (4): 228–33.

———. 1995. Cognitive ethology: The comparative study of animal minds. In *Blackwell Companion to Cognitive Science*, ed. W. Bechtel and G. Graham. Oxford: Blackwell.

———. 2000a. Animal emotions: Exploring passionate natures. *BioScience* 50 (10): 861–70.

———. 2000b. *The Smile of a Dolphin: Remarkable Accounts of Animal Emotions*. New York: Discovery Books.

Bel'kovich, V. M. 1991. Herd structure, hunting, and play: Bottlenose dolphins in the Black Sea. In *Dolphin Societies: Discoveries and Puzzles*, ed. K. Pryor and K. S. Norris, 17–78. Berkeley: University of California Press.

Berta, A., and J. L. Sumich. 2004. *Marine Mammals: Evolutionary Biology*. San Diego: Academic.

Biro, D., N. Inoue-Nakamura, R. Tonooka, G. Yamakoshi, C. Sousa, and T. Matsuzawa. 2003. Cultural innovation and transmission of tool use in wild chimpanzees: Evidence from field experiments. *Animal Cognition* 6 (4): 213–23.

Bowles, A. E., W. G. Young, and E. D. Asper. 1988. Ontogeny of stereotyped calling of a killer whale calf, *Orcinus orca*, during her first year. *Rit Fiskideildar* 11: 251–75.

Box, H. O. 1984. *Primate Behaviour and Social Ecology*. New York: Chapman and Hall.

Bradbury, J. W., and S. L. Vehrencamp. 1998. *Principles of Animal Communication*. Sunderland, MA: Sinauer.

Bradshaw, G. A., and A. N. Schore. 2007. How elephants are opening doors: Neuroethology, attachment, and social context. *Ethology* 113: 426–36.

Bradshaw, G. A., A. N. Schore, J. L. Brown, J. H. Poole, and C. J. Moss. 2005. Elephant breakdown. *Nature* 433: 807.

Brensing, K., K. Linke, and D. Todt. 2003. Can dolphins heal by ultrasound? *Journal of Theoretical Biology* 225: 99–105.

Buck, C. D., and J. P. Schroeder. 1990. Public health significance of marine mammal disease. In *CRC Handbook of Marine Mammal Medicine: Health, Disease and Rehabilitation*, ed. L. A. Dierauf, 163–73. Boston: CRC.

Caldwell, D. K., and M. C. Caldwell. 1977. Cetaceans. In *How Animals Communicate*, ed. T. A. Sebeok, 794–808. Bloomington: Indiana University Press.

Caldwell, M. C., and D. K. Caldwell. 1965. Individualized whistle contours in bottle-nosed dolphins (*Tursiops truncatus*). *Nature* 207: 434–35.

———. 1968. Vocalizations of naive captive dolphins in small groups. *Science* 159: 1121–23.

Caldwell, M. C., D. K. Caldwell, and J. B. Siebanaler. 1965. Observations on captive and wild Atlantic bottlenosed dolphins, *Tursiops truncatus*, in the southeastern Gulf of Mexico. *Los Angeles County Museum Contributions in Science* 91: 1–10.

Caldwell, M. C., D. K. Caldwell, and P. L. Tyack. 1990. Review of the signature-whistle hypothesis for the Atlantic bottlenose dolphin. In *The Bottlenose Dolphin*, ed. S. Leatherwood and R. R. Reeves, 199–234. San Diego: Academic.

Carlson, C. and T. Frohoff. In preparation. Viewing cetaceans: Comparing whale watching with captivity. Forthcoming in *Marine Mammals of Costa Rica*, ed. J. Rodríguez-Fonseca, A. Acevedo-Gutiérrez, and P. Cubero-Pardo.

Caro, T. M., and M. D. Hauser. 1992. Is there teaching in nonhuman animals? *Quarterly Review of Biology* 67: 151–74.

Carter, N. 1982. Effects of psycho-physiological stress on captive dolphins. *International Journal for the Study of Animal Problems* 3 (3): 193–98.

Christal, J., H. Whitehead, and E. Lettevall. 1998. Sperm whale social units: Variation and change. *Canadian Journal of Zoology* 76: 1431–40.

Cipriano, F. W. 1992. Behavior and occurrence patterns, feeding ecology, and life history of dusky dolphins (*Lagenorhynchus obscurus*) off Kaikoura, New Zealand. Ph.D. diss., University of Arizona.

Clark, C. W. 1983. Acoustic communication and behavior of the southern right whale (*Eubalaena australis*). In *Communication and Behavior of Whales*, ed. R. Payne, 163–98. Boulder, CO: Westview.

Clark, C. W., and M. Mangel. 1986. The evolutionary advantages of group foraging. *Theoretical Population Biology* 30: 45–75.

Connor, R. C. 1990a. Alliances among male bottlenose dolphins and a comparative analysis of mutualism. Ph.D. diss., University of Michigan.

———. 1990b. Dolphin alliances and coalitions. In *Coalitions and Alliances in Humans and Other Animals*, ed. A. H. Harcourt and F. de Waal, 415–43. Oxford: Oxford University Press.

Connor, R. C., and M. Krützen. 2003. Levels and patterns in dolphin alliance formation. In *Animal Social Complexity: Intelligence, Culture, and Individualized Societies*, ed. F. B. M. de Waal and P. L. Tyack, 115–20. Cambridge, MA: Harvard University Press.

Connor, R. C., and J. Mann. 2006. Social cognition in the wild: Machiavellian dolphins? In *Rational Animals?* ed. S. Hurley and M. Nudds, 329–67. Oxford: Oxford University Press.

Connor, R. C., and K. S. Norris. 1982. Are dolphins reciprocal altruists? *American Naturalist* 119 (3): 358–74.

Connor, R. C., and R. A. Smolker. 1985. Habituated dolphins (*Tursiops* sp.) in Western Australia. *Journal of Mammalogy* 66: 398–400.

———. 1996. "Pop" goes the dolphin: A vocalization male bottlenose dolphins produce during consortships. *Behaviour* 133: 643–62.

Connor, R. C., R. M. Heithaus, and L. M. Barre. 1999. Superalliance of bottlenose dolphins. *Nature* 571: 571–72.

———. 2001. Complex social structure, alliance stability and mating access in a bottlenose dolphin "super-alliance." *Proceedings of the Royal Society B: Biological Sciences* 268: 263–67.

Connor, R. C., J. Mann, P. L. Tyack, and H. Whitehead. 1998. Social evolution in toothed whales. *Trends in Ecology and Evolutionary Biology* 13 (6): 228–32.

Connor, R. C., R. M. Heithaus, P. Berggren, and J. L. Miksis. 2000. "Kerplunking": Surface fluke-splashes during shallow-water bottom foraging by bottlenose dolphins. *Marine Mammal Science* 16: 646–53.

Connor, R. C., R. A. Smolker, and A. F. Richards. 1992. Two levels of alliance formation among male bottlenose dolphins (*Tursiops* sp.). *Proceedings of the National Academy of Sciences of the United States* 89 (3): 987–90.

Connor, R. C., R. Smolker, and L. Bejder. 2006. Synchrony, social behaviour and alliance affiliation in Indian Ocean bottlenose dolphins, *Tursiops aduncus. Animal Behaviour* 72 (6): 1371–78.

Constantine, R. 1995. Monitoring the commercial swim-with-dolphin operations with the bottlenose (*Tursiops truncatus*) and common dolphins (*Delphinus delphis*) in the Bay of Islands, New Zealand. M.A. thesis, University of Auckland.

———. 2001. Increased avoidance of swimmers by bottlenose dolphins (*Tursiops truncatus*) due to long-term exposure to swim-with-the-dolphin tourism. *Marine Mammal Science* 17 (4): 689–702.

Couquiaud, L. 2005. A survey of the environments of cetaceans in human care. *Aquatic Mammals* 31 (3): 279–385.

Courbis, S. 2004. Behavior of Hawai'ian spinner dolphins (*Stenella longirostris*) in response to vessels/swimmers. M.A. thesis, San Francisco State University.

Cranford, T. W., M. Amundin, and K. S. Norris. 1992. Functional morphology and homology in the odontocete nasal complex: Implications for sound generation. *Journal of Morphology* 228: 223–85.

Darwin, C. 1936 [1871]. *The Descent of Man and Selection in Relation to Sex.* New York: Random House.

———. 1965. *The Expression of the Emotions in Man and Animals.* Chicago: University of Chicago Press.

Dawkins, M. S. 1980. *Animal Suffering: The Science of Animal Welfare.* London: Chapman and Hall.

———. 1990. From an animal's point of view: Motivation, fitness, and animal welfare. *Behavior and Brain Science* 13: 1–61.

Dawson, S. M. 1991. Clicks and communication: The behavioural and social contexts of Hector's dolphin vocalizations. *Ethology* 88: 265–76.

Deecke, V. B. 2006. Studying marine mammal cognition in the wild: A review of four decades of playback experiments. *Aquatic Mammals* 32 (4): 461–82.

Deecke, V. B., J. K. Ford, and P. Spong. 2000. Dialect change in resident killer whales: Implications for vocal learning and cultural transmission. *Animal Behaviour* 60 (5): 629–38.

Defran, R. H., and K. Pryor. 1980. The behavior and training of cetaceans in captivity. In *Cetacean Behavior: Mechanisms and Functions*, ed. L. M. Herman, 319–62. New York: John Wiley and Sons.

Dennett, D. C. 1991. *Consciousness Explained*. Boston: Little, Brown.

Dere, E., E. Kart-Teke, J. P. Huston, and M. A. De Souza Silva. 2006. The case for episodic memory in animals. *Neuroscience and Biobehavioral Reviews* 30 (8): 1206–24.

Doak, W. 1981. *Dolphin, Dolphin*. Auckland, New Zealand: Hodder and Stoughton.

———. 1988. *Encounters with Whales and Dolphins*. Dobbs Ferry, NY: Sheridan House.

Domning, D. P. 1991. A former dolphin-human fishing cooperative in Australia. *Marine Mammal Science* 7 (1): 94–96.

Donaldson, F. 1982. Play and contest in human-animal relationships. *Lomi School Bulletin*, Fall, pp. 5–9.

Douglas-Hamilton, I., S. Bhalla, G. Wittemyer, and F. Vollrath. 2006. Behavioural reactions of elephants towards a dying and deceased matriarch. *Applied Animal Behaviour Science* 100: 87–102.

Dreher, J. J. 1961. Linguistic considerations of porpoise sounds. *Journal of the Acoustical Society of America* 33: 1799–800.

Dreher, J. J., and W. E. Evans. 1964. Cetacean communication. In *Marine Bioacoustics*, vol. 1, ed. W. N. Tavolga, 373–99. Oxford: Pergamon.

Dudzinski, K. M. 1996. Communication in Atlantic spotted dolphins (*Stenella frontalis*): Relationships between vocal and behavioral activities. Ph.D. diss., Texas A&M University.

———. 1998. Contact behavior and signal exchange among Atlantic spotted dolphins (*Stenella frontalis*). *Aquatic Mammals* 24 (3): 129–42.

———. 2003. Letting dolphins speak: Are we listening? In *Between Species: Celebrating the Dolphin-Human Bond*, ed. T. Frohoff and B. Peterson, 286–95. San Francisco: Sierra Club Books.

Dudzinski, K. M., and D. Newborough. 1997. Concurrent recording of dolphin behaviors, frequency-modulated tones, and pulsed vocalizations (including echolocation clicks) underwater with a swimmer-propelled system. Proceedings of the Underwater Bio-Sonar Systems and Bioacoustics Symposium, Underwater Acoustics Group, December 16–17, 1997, Loughborough, England.

Dudzinski, K. M., C. W. Clark, and B. Würsig. 1995. A mobile video/acoustic system for simultaneously recording dolphin behavior and vocalizations underwater. *Aquatic Mammals* 21 (3): 187–93.

Dudzinski, K. M., E. Douaze, and J. A. Thomas. 2002. Communication. In *Encyclopedia of Marine Mammals*, ed. W. F. Perrin, B. Würsig, and H. C. M. Thewissen, 248–68. San Diego: Academic.

Dudzinski, K. M., T. G. Frohoff, and N. L. Crane. 1995. Occurrence and behavior of a lone, sociable female dolphin (*Tursiops truncatus*) off the coast of Belize. *Aquatic Mammals* 21 (2): 149–53.

Dudzinski, K. M., S. A. Kuczaj, C. A. Ribic, and J. D. Gregg. Submitted. Flipper's flipper: A comparison of how, where, and why spotted and bottlenose dolphins use their pectoral fins to touch peers. Forthcoming in *Animal Behaviour*.

Dudzinski, K. M., M. Sakai, K. Masaki, K. Kogi, T. Hishii, and M. Kurimoto. 2003. Behavioural observations of bottlenose dolphins towards two dead conspecifics. *Aquatic Mammals* 29 (1): 108–16.

Duffus, D. A., and P. Deardon. 1992. Whales, science, and protected area management in British Columbia, Canada. *George Wright Forum* 9: 79–87.

Evans, P. G. H. 1987. *The Natural History of Whales and Dolphins*. New York: Facts on File.

Fagen, R. 1981. *Animal Play Behavior*. New York: Oxford University Press.

Fedorowicz, S. M., D. A. Beard, and R. C. Connor. 2003. Food sharing in wild bottlenose dolphins. *Aquatic Mammals* 29 (3): 355–59.

Fellner, W., G. B. Bauer, and H. E. Harley. 2006. Cognitive implications of synchrony in dolphins: A review. *Aquatic Mammals* 32 (4): 511–16.

Fertl, D. 1994. Occurrence patterns and behavior of bottlenose dolphins (*Tursiops truncatus*) in the Galveston ship channel, Texas. *Texas Journal of Science* 46 (4): 299–317.

Fertl, D., and B. Würsig. 1995. Coordinated feeding by Atlantic spotted dolphins (*Stenella frontalis*) in the Gulf of Mexico. *Aquatic Mammals* 21: 3–5.

Finley, K. J., G. W. Miller, R. A. Davis, and C. R. Greene. 1990. Reactions of belugas, *Delphinaptera leucas*, and narwals, *Monodon monoceros*, to ice-breaking ships of the Canadian high arctic. *Canadian Bulletin of Fisheries and Aquatic Sciences* 224: 97–117.

Ford, J. K. B. 1989. Acoustic behaviour of resident killer whales (*Orcinus orca*) off Vancouver Island, British Columbia, Canada. *Canadian Journal of Zoology* 67: 727–45.

———. 1991. Vocal traditions among resident killer whales (*Orcinus orca*) in coastal waters of British Columbia. *Canadian Journal of Zoology* 69: 1454–83.

Ford, J. K. B., G. M. Ellis, L. G. Barrett-Lennard, A. B. Morton, R. S. Palm, and K. C. Balcomb III. 1999. Dietary specialization in two sympatric populations of killer whales (*Orcinus orca*) in coastal British Columbia and adjacent waters. *Canadian Journal of Zoology* 76: 1456–71.

Forest, A. M. 2001. The Hawai'ian spinner dolphin, *Stenella longirostris:* Effects of tourism. M.A. thesis, Texas A&M University.

Fossey, D. 1983. *Gorillas in the Mist.* Boston: Houghton Mifflin.

Fraser, J., D. Reiss, P. Boyle, K. Lemcke, J. Sickler, and E. Elliott. 2006. Dolphins in popular literature and media. *Society and Animals* 14 (4): 321–49.

Frohoff, T. G. 1993. Behavior of captive bottlenose dolphins (*Tursiops truncatus*) and humans during controlled in-water interactions. M.A. thesis, Texas A&M University.

———. 1996. Behavior of bottlenose (*Tursiops truncatus*) and spotted dolphins (*Stenella frontalis*) relative to human interaction. Ph.D. diss., Union Institute.

———. 1998. Beyond species. In *Intimate Nature*, ed. B. Peterson, L. Hogan, and D. Metzger, 78–84. New York: Ballantine.

———. 2000a. The dolphin's smile. In *The Smile of a Dolphin: The Emotional Lives of Animals*, ed. M. Bekoff, 72–73. New York: Discovery Books.

———. 2000b. Behavioral indicators of stress in odontocetes during interactions with humans: A preliminary review and discussion. Report to the International Whaling Commission Scientific Committee, SC/52/WW2.

———. 2003. The kindred wild. In *Between Species: Celebrating the Dolphin-Human Bond*, ed. T. Frohoff and B. Peterson, 56–71. San Francisco: Sierra Club Books.

———. 2004. Stress in dolphins. In *Encyclopedia of Animal Behavior*, ed. M. Bekoff, 1158–64. Westport, CT: Greenwood.

———. 2007. Dolphins and humans: From the sea to the swimming pool. In *Encyclopedia of Human-Animal Relationships*, ed. Marc Bekoff, 1096–106. Westport, CT: Greenwood.

———. In press. Marine animal welfare. In *Encyclopedia of Tourism and Recreation in Marine Environments*, ed. M. Lück. Wallingford, UK: CABI.

Frohoff, T. G., B. A. Bilgre, A. M. Sanders, and K. M. Dudzinski. 1996. Prediction and management of high-risk behavior in a lone, sociable dolphin. In Proceedings from the Twenty-First International Meeting for the Study of Marine Mammals, April 8–12, 1996, Chetumal, Mexico, p. 9.

Frohoff, T. G., and J. M. Packard. 1995. Human interactions with free-ranging and captive bottlenose dolphins. *Anthrozoös* 8 (1): 44–54.

Frohoff, T. G., J. M. Packard, and R. H. Benson. 1996. Variation in whistle type and rate produced by captive dolphins relative to in-water interactions with humans. *Journal of the Acoustical Society of America* 99 (4): 2558.

Frohoff, T., and B. Peterson. 2003. *Between Species: Celebrating the Dolphin-Human Bond.* San Francisco: Sierra Club Books.

Frohoff, T. G., C. S. Vail, and M. Bossley, eds. 2005. Unpublished book for the workshop on research and management of solitary, sociable odontocetes, Sixteenth Biennial Conference on the Biology of Marine Mammals, December 10, 2005, San Diego.

———. 2006. Preliminary proceedings of the workshop on the research and management of solitary, sociable odontocetes. Sixteenth Biennial Conference on the Biology of Marine Mammals, December 10, 2005, San Diego.

Gales, N., M. Hindell, and R. Kirkwood, eds. 2003. *Marine Mammals: Fisheries, Tourism and Management Issues.* Collingwood, Australia: CISRO.

Gallup, G. G. 1970. Chimpanzees: Self-recognition. *Science* 167: 86–87.

Gardner, R. A., and B. T. Gardner. 1969. Teaching sign language to a chimpanzee. *Science* 165: 664–72.

Gaskin, D. E. 1982. *The Ecology of Whales and Dolphins.* Exeter, NH: Heinemann.

Gazda, S. K., R. C. Connor, R. K. Edgar, and F. Cox. 2005. A division of labour with role specialization in group-hunting bottlenose dolphins (*Tursiops truncatus*) off Cedar Key, Florida. *Proceedings of the Royal Society B: Biological Sciences* 272 (1559): 135–40.

Geraci, J. R. 1986. Marine mammals: Husbandry. In *Zoo and Wild Animal Medicine*, ed. M. E. Fowler, 757–60. Philadelphia: Saunders.

Goodall, J. 1971. *In the Shadow of Man.* Boston: Houghton Mifflin.

———. 1986. *The Chimpanzees of Gombe: Patterns of Behavior.* Cambridge, MA: Belknap Press of Harvard University Press.

Greenwood, A. G. 1977. A stereotyped behaviour pattern in dolphins. *Aquatic Mammals* 6 (1): 15–17.

Gregg, J. D., K. M. Dudzinski, and H. V. Smith. 2007. A method for estimating relative head angle and spatial distance of dolphins from underwater video footage. *Animal Behaviour* 75 (3): 1181–86.

Grier, J. W., and T. Burk. 1992. *Biology of Animal Behavior.* Saint Louis: Mosby-Year Book.

Griffin, D. R. 1986. *Listening in the Dark.* Ithaca, NY: Cornell University Press.

Gubbins, C. M. 2002. *The Dolphins of Hilton Head: Their Natural History.* Columbia: University of South Carolina Press.

Guinet, C. 1992. Comportement de chasse des orques (*Orcinus orca*) autour des îles Crozet [Hunting behavior in killer whales (*Orcinus orca*) around the Crozet Islands]. *Canadian Journal of Zoology* 70: 1656–67.

Guinet, C., and J. Bouvier. 1995. Development of intentional stranding hunting techniques in killer whale (*Orcinus orca*) calves at Crozet Archipelago. *Canadian Journal of Zoology* 73: 27–33.

Hankins, T. N. 1993. Interactions and play between captive bottlenose dolphins and humans in a "swim-with-the-dolphins" program. M.A. thesis, Florida International University.

Harcourt, A. H., and F. B. M. de Waal. 1992. *Coalitions and Alliances in Humans and Other Animals*. New York: Oxford University Press.

Harley, H. E., H. L. Roitblat, and P. E. Nachtigall. 1996. Object representation in the bottlenose dolphin (*Tursiops truncatus*): Integration of visual and echoic information. *Journal of Experimental Psychology: Animal Behavior Processes* 22 (2): 164–74.

Hart, D., and J. W. Whitlow, Jr. 1995. The experience of self in the bottlenose dolphin. *Conscious Cognition* 4 (2): 244–47.

Hauser, M. D. 1996. *The Evolution of Communication*. Cambridge, MA: MIT Press.

Hediger, H. 1955. *Studies of the Psychology and Behavior of Captive Animals in Zoos and Circuses*. New York: Criterion.

———. 1964. *Wild Animals in Captivity*. New York: Dover.

Heimlich-Boran, J. R. 1988. Behavioral ecology of killer whales (*Orcinus orca*) in the Pacific Northwest. *Canadian Journal of Zoology* 66: 565–78.

Hemila, S., and S. Nummelaa. 1999. A model of the odontocete middle ear. *Hearing Research* 133: 82–97.

Hendry, J. L. 2003. The ontogeny of echolocation in the Atlantic bottlenose dolphin (*Tursiops truncatus*). Ph.D. diss., University of Southern Mississippi.

Herman, L. M. 1980. Cognitive characteristics of dolphins. In *Cetacean Behavior: Mechanisms and Functions*, ed. L. M. Herman, 363–430. New York: John Wiley and Sons.

———. 1986. Cognition and language competencies of bottlenosed dolphins. In *Dolphin Cognition and Behavior: A Comparative Approach*, ed. J. A. Schusterman, J. A. Thomas, and F. G. Wood, 221–52. Hillsdale, NJ: Lawrence Erlbaum Associates.

———. 2006. Intelligence and rational behaviour in the bottlenosed dolphin. In *Rational Animals?* ed. S. Hurley and M. Nudds, 439–67. Oxford: Oxford University Press.

Herman, L. M., and W. N. Tavolga. 1980. The communication systems of cetaceans. In *Cetacean Behavior: Mechanisms and Functions*, ed. L. M. Herman, 149–209. New York: John Wiley and Sons.

Herman, L. M., S. L. Abichandani, A. N. Elhajj, E. Y. K. Herman, J. L. Sanchez, and A. A. Pack. 1999. Dolphins (*Tursiops truncatus*) comprehend the referential character of the human pointing gesture. *Journal of Comparative Psychology* 113 (4): 347–64.

Herman, L. M., D. S. Matus, E. Y. Herman, M. Ivancic, and A. A. Pack. 2001. The bottlenosed dolphin's (*Tursiops truncatus*) understanding of gestures as symbolic representations of its body parts. *Animal Learning and Behavior* 29 (3): 250–64.

Herman, L. M., P. Morrel-Samuels, and A. A. Pack. 1990. Bottlenosed dolphin and human recognition of veridical and degraded video displays of an artificial gestural language. *Journal of Experimental Psychology, General* 119 (2): 215–30.

Herman, L. M., M. F. Peacock, M. P. Yunker, and C. J. Madsen. 1975. Bottlenosed dolphin: Double-slit pupil yields equivalent aerial and underwater diurnal acuity. *Science* 189: 650–52.

Herman, L. M., D. G. Richards, and J. P. Wolz. 1984. Comprehension of sentences by bottle-nosed dolphins. *Cognition* 16: 129–219.

Herzing, D. L. 1990. Underwater and close up with spotted dolphins. *Whalewatcher (Journal of the American Cetacean Society)* 24: 16–19.

———. 1991a. Family, friends and neighbors. *Whalewatcher (Journal of the American Cetacean Society)* 26: 13–15.

———. 1991b. Underwater behavioral observations and sound correlations of free-ranging Atlantic spotted dolphin, *Stenella frontalis*, and bottlenose dolphin, *Tursiops truncatus*. Abstracts of the Ninth Biennial Conference on the Biology of Marine Mammals, December 5–9, 1991, Chicago.

———. 1993. Dolphins in the wild: An eight year field study on dolphin communication and interspecies interaction. Ph.D. diss., Union Institute.

———. 2000. Acoustics and social behavior of wild dolphins: Implications for a sound society. In *Hearing by Whales and Dolphins*, ed. W. W. L. Au, A. N. Popper, and R. R. Fay, 225–72. New York: Springer-Verlag.

Hof, P. R., I. I. Glezer, E. A. Nimchinsky, and J. M. Erwin. 2000. Neurochemical and cellular specializations in the mammalian neocortex reflect phylogenetic relationships: Evidence from primates, cetaceans, and artiodactyls. *Brain Behavior and Evolution* 55: 300–310.

Hof, P., R. Chanis, and L. Marino. 2005. Cortical complexity in cetacean brains. *Anatomical Record* 287A: 1142–52.

Hoyt, E. 2002. Whale watching. In *Encyclopedia of Marine Mammals*, ed. W. F. Perrin, B. Würsig, and J. G. M. Thewissen, 1305–10. San Diego: Academic.

———. 2005. *Marine Protected Areas for Whales, Dolphins and Porpoises: A World Handbook for Cetacean Habitat Conservation*. London: Earthscan.

Hughes, B. O., and I. J. H. Duncan. 1988. The notion of ethological "need," models of moti-vation and animal welfare. *Animal Behaviour* 36: 1696–1707.

Humphries, T. 2003. Effectiveness of dolphin-assisted therapy as a behavioral intervention for young children with disabilities. *Bridges* 1 (6): 1–9.

Hurley, S., and M. Nudds, eds. 2006. *Rational Animals?* Oxford: Oxford University Press.

Janik, V. M. 2000. Whistle matching in wild bottlenose dolphin (*Tursiops truncatus*). *Science* 289: 1355–57.

Janik, V. M., and P. J. B. Slater. 1998. Context-specific use suggests that bottlenose dolphin signature whistles are cohesion calls. *Animal Behaviour* 56: 829–38.

———. 2000. The different roles of social learning in vocal communication. *Animal Behaviour* 60: 1–11.

Jerison, H. J. 1973. *Evolution of the Brain and Intelligence.* New York: Academic.

Johnson, C. M., and K. S. Norris. 1986. Delphinid social organization and social behavior. In *Dolphin Cognition and Behavior: A Comparative Approach,* ed. R. J. Schusterman, J. A. Thomas, and F. G. Wood, 335–46. Hillsdale, NJ: Lawrence Erlbaum Associates.

Karczmarski, L., M. Thornton, and V. G. Cockcroft. 1997. Description of selected behaviours of humpbacked dolphins, *Sousa chinensis. Aquatic Mammals* 23: 127–33.

Kellert, S. R. 1996. *The Value of Life: Biological Diversity and Human Society.* Washington, DC: Island Press.

———. 1999. American perceptions of marine mammals and their management. Washington, DC: Humane Society of the United States.

Ketten, D. R. 1992. The marine mammal ear: Specializations for aquatic audition and echo-location. In *The Evolutionary Biology of Hearing,* ed. D. Webster, R. R. Fay, and A. N. Popper, 717–54. New York: Springer-Verlag.

Kiszka, J. J. 2007. Atypical associations between dugongs (*Dugong dugon*) and dolphins in a tropical lagoon. *Journal of the Marine Biological Association of the UK* 87: 101–4.

Krebs, J. R., and R. Dawkins. 1984. Animal signals: Mind-reading and manipulation. In *Behavioral Ecology: An Evolutionary Approach,* 2nd ed., ed. J. R. Krebs and N. B. Davies, 380–402. Boston: Blackwell Scientific.

Kruse, S. 1991. The interactions between killer whales and boats in Johnstone Strait, B.C. In *Dolphin Societies: Discoveries and Puzzles,* ed. K. Pryor and K. S. Norris, 149–60. Berkeley: University of California Press.

Krützen, M., J. Mann, M. R. Heithaus, R. C. Connor, L. Bejder, and W. B. Sherwin. 2005. Cultural transmission of tool use in bottlenose dolphins. *Proceedings of the National Academy of Sciences of the United States* 102 (25): 8939–43.

Kuczaj, S. A., and L. E. Highfill. 2005. Dolphin play: Evidence for cooperation and culture? *Behavioral Brain Science* 28 (5): 705–6.

Kuczaj, S. A., II, and D. B. Yeater. 2006. Dolphin imitation: Who, what, when, and why? *Aquatic Mammals* 32 (4): 413–22.

Kudo, H., and R. I. M. Dunbar. 2001. Neocortex size and social network size in primates. *Animal Behaviour* 62: 711–22.

Kuznetsov, V. B. 1990. Chemical sense in dolphins: Quasiolfaction. In *Sensory Methods of Cetaceans: Laboratory and Field Evidence*, ed. J. Thomas and R. Kastelein, 481–502. New York: Plenum.

Kyngdon, D. J., E. O. Minot, and K. J. Stafford. 2003. Behavioural responses of captive common dolphins *Delphinus delphis* to a "Swim-with-Dolphin" programme. *Applied Animal Behaviour Science* 81: 163–170.

Lammers, M. O., and W. W. L. Au. 2003. Directionality in the whistles of Hawaiian spinner dolphins *Stenella longirostris:* A signal feature to cue direction of movement? *Marine Mammal Science* 19: 249–64.

Langbauer, W. R., Jr., K. B. Payne, R. A. Charif, and E. M. Thomas. 1989. Responses of captive African elephants to playback of low-frequency calls. *Canadian Journal of Zoology* 67: 2604–7.

Lautin, A. 2001. *The Limbic Brain.* New York: Kluwer Academic/Plenum.

Leatherwood, S., and R. R. Reeves. 1983. *The Sierra Club Handbook of Whales and Dolphins.* San Francisco: Sierra Club Books.

Leavens, D. A. 2004. Manual deixis in apes and humans. *Interaction Studies* 5: 387–408.

Lilly, J. C. M. 1961. *Man and Dolphin.* Garden City, NY: Doubleday.

———. 1967. *The Mind of the Dolphin.* Garden City, NY: Doubleday.

———. 1973. *The Centre of the Cyclone: An Autobiography of Inner Space.* London: Calders and Boyars.

———. 1975. *Lilly on Dolphins.* Garden City, NY: Anchor.

———. 1978 [1961]. *Communication between Man and Dolphin: The Possibilities of Talking with Other Species.* New York: Crown.

Lilly, J. C., and A. Montagu. 1963. *The Dolphin in History.* Chicago: William Andrews Clark Memorial Library.

Lockyer, C. 1990. Review of incidents involving wild, sociable dolphins, worldwide. In *The Bottlenose Dolphin*, ed. S. Leatherwood and R. R. Reeves, 337–53. San Diego: Academic.

Lusseau, D. 2003. How do tour boats affect the behavioral state of bottlenose dolphins (*Tursiops* sp.): Applying Markov chain modeling to the field study of behavior. *Conservation Biology* 17: 1785–93.

Lusseau, D., and M. E. Newman. 2004. Identifying the role that animals play in their social networks. *Proceedings of the Royal Society B: Biological Sciences* 271 Suppl. 6: S477–81.

MacLeon, P. D. 1970. *The Triune Brain in Evolution: Role in Paleocerebral Functions.* New York: Plenum.

Manger, P. R. 2006. An examination of cetacean brain structure with a novel hypothesis cor-
relating thermogenesis to the evolution of a big brain. *Biological Reviews of the Cambridge
Philosophical Society* 81 (2): 293–338.

Mann, J., and B. Smuts. 1999. Behavioral development of wild bottlenose dolphin newborns.
Behaviour 136: 529–66.

Mann, J., R. C. Connor, P. L. Tyack, and H. Whitehead. 2000. *Cetacean Societies Field Studies
of Dolphins and Whales.* Chicago: University of Chicago Press.

Marine Mammal Commission. 2007. *Marine Mammals and Noise: A Sound Approach to Research
and Management.* A report to Congress from the Marine Mammal Commission. March.

Marino, L. 2002. Convergence of complex cognitive abilities in cetaceans and primates.
Brain and Behavioral Evolution 59 (1–2): 21–32.

———. 2004. Cetacean brain evolution: Multiplication generates complexity. *International
Journal of Comparative Psychology* 17: 1–16.

Marino, L., and S. O. Lilienfeld. 1998. Dolphin assisted therapy: Flawed data and flawed
conclusions. *Anthrozoös* 11 (4): 194–200.

———. 2007. Dolphin assisted therapy: More flawed data and more flawed conclusions.
Anthrozoös 20 (3): 239–49.

Marino, L., D. W. McShea, and M. D. Uhen. 2004. Origin and evolution of large brains in
toothed whales. *Anatomical Record* 281A (2): 1247–55.

Marino, L., J. K. Rilling, S. K. Lin, and S. H. Ridgway. 2000. Relative volume of the cerebel-
lum in dolphins and comparison with anthropoid primates. *Brain Behavior and Evolution* 56
(4): 204–11.

Marten, K., and S. Psarakos. 1995. Using self-view television to distinguish between self-
examination and social behavior in the bottlenose dolphin (*Tursiops truncatus*). *Consciousness
and Cognition* 4: 205–24.

Marten, K., K. Shariff, S. Psarakos, and D. J. White. 1996. Ring bubbles of dolphins. *Scientific
American* 275 (2): 82–87.

Martin, P., and P. Bateson. 1986. *Measuring Behavior: An Introductory Guide.* New York:
Cambridge University Press.

Mason, W. A. 1999. Experiential influences on the development of expressive behaviors in
rhesus monkeys. In *The Development of Expressive Behavior: Biology-Environment Interactions,*
ed. G. Zivin, 117–51. New York: Academic.

Mass, A. M., and A. Y. Supin. 2002. Vision. In *Encyclopedia of Marine Mammals,* ed. W. F.
Perrin, B. Würsig, and J. G. M. Thewissen, 1280–93. San Diego: Academic.

Mass, A. M., and S. A. Ya. 1995. Ganglion cell topography of the retina in the bottlenosed
dolphin, *Tursiops truncatus. Brain Behavior and Evolution* 45 (5): 257–65.

Mazet, J. A., T. D. Hunt, and M. H. Ziccardi. 2004. Assessment of the risk of zoonotic disease transmission to marine mammal workers and the public: Survey of occupational risks. Final Report prepared for United States Marine Mammal Commission, RA No. K005486-01.

McBride, A. F., and D. O. Hebb. 1948. Behavior of the captive bottlenose dolphin, *Tursiops truncatus*. *Journal of Comparative Physiology* 41: 111–23.

McCowan, B., L. Marino, E. Vance, L. Walke, and D. Reiss. 2000. Bubble ring play of bottlenose dolphins (*Tursiops truncatus*): Implications for cognition. *Journal of Comparative Psychology* 114 (1): 98–106.

McCowan, B., and D. Reiss. 1995a. Maternal aggressive contact vocalizations in captive bottlenose dolphins (*Tursiops truncatus*): Wide-band, low-frequency signals during mother/aunt-infant interactions. *Zoo Biology* 14: 293–309.

———. 1995b. Quantitative comparison of whistle repertoires from captive adult bottlenose dolphins (Delphinidae, *Tursiops truncatus*): A re-evaluation of the signature whistle hypothesis. *Journal of Ethology* 100: 194–209.

———. 1995c. Whistle contour development in captive-born infant bottlenose dolphins (*Tursiops truncatus*): Role of learning. *Journal of Comparative Psychology* 109 (3): 242–60.

Melillo, K. 2005. Spotted and bottlenose dolphin interactions around Bimini. Video presentation at the Sixteenth Biennial Conference on the Biology of Marine Mammals, December 12–16, 2005, San Diego.

Mercado, E., III, R. K. Uyeyama, A. A. Pack, and L. M. Herman. 1999. Memory for action events in the bottlenosed dolphin. *Animal Cognition* 2: 17–25.

Meyer-Holzapfel, M. 1968. Abnormal behavior in zoo animals. In *Abnormal Behavior in Animals*, ed. M. W. Fox, 476–503. Philadelphia: W. B. Saunders.

Miklósi, Ã., R. Polgárdi, J. Topál, and V. Csányi. 1998. Use of experimenter-given cues in dogs. *Animal Cognition* 1 (2): 113–21.

Miller, G. A. 2003. The cognitive revolution: A historical perspective. *Trends in Cognitive Sciences* 7 (3): 141–44.

Moehlman, P. D. 1989. Intraspecific variation in canid social systems. In *Carnivore Behavior, Ecology, and Evolution*, ed. J. L. Gittleman, 143–63. Ithaca, NY: Cornell University Press.

Morgan, S. M. 1968. The sagacious dolphin. *Natural History* 77: 32–39.

Morisaka, T., M. Shinohara, F. Nakahara, and T. Akamatsu. 2005. Geographic variations in the whistles among three Indo-Pacific bottlenose dolphin *Tursiops aduncus* populations in Japan. *Fisheries Science* 71: 568–76.

Morton, A. B., and H. K. Symonds. 2002. Displacement of *Orcinus orca* (L.) by high amplitude sound in British Columbia, Canada. ICES *Journal of Marine Science* 59: 71–80.

Morton, D. B., and P. H. M. Griffiths. 1985. Guidelines on the recognition of pain and discomfort in experimental animals and an hypothesis for assessment. *Veterinary Record* 116: 431–36.

Morton, E. S. 1977. On the occurrence and significance of motivation-structural rules in some bird and mammal sounds. *American Naturalist* 111 (981): 855–69.

Moss, C. 1982. *Portraits in the Wild Animal Behavior in East Africa*, 2nd ed. Chicago: University of Chicago Press.

Nachtigal, P. 1986. Vision, audition, and chemoreception in dolphins and other marine mammals. In *Dolphin Cognition and Behavior: A Comparative Approach*, ed. R. J. Schusterman, J. A. Thomas, and F. G. Wood, 79–113. Hillsdale, NJ: Lawrence Erlbaum Associates.

National Marine Fisheries Service. 1990. A final environmental impact statement on the use of marine mammals in "swim-with-the-dolphin" programs. Office of Protected Resources, Silver Spring, MD.

Nikaido, M., A. P. Rooney, and N. Okada. 1999. Phylogenic relationships among cetartiodactyls based on insertions of short and long interspersed elements: Hippopotamuses are the closest extant relatives of whales. *Proceedings of the National Academy of Sciences of the United States* 96: 10261–66.

Nollman, J. 1987. *Dolphin Dreamtime*. New York: Bantam.

Norris, K. S. 1969. The echolocation of marine mammals. In *The Biology of Marine Mammals*, ed. H. T. Andersen, 425–75. San Diego: Academic.

Norris, K. S., and T. P. Dohl. 1980. The structure and function of cetacean schools. In *Cetacean Behavior: Mechanisms and Functions*, ed. L. M. Herman, 211–62. New York: John Wiley and Sons.

Norris, K. S., and J. H. Prescott. 1961. Observations on Pacific cetaceans of Californian and Mexican waters. *University of California Publications in Zoology* 63: 135–50.

Norris, K. S., and C. R. Schilt. 1988. Cooperative societies in three-dimensional space: On the origins of aggregations, flocks, and schools, with special reference to dolphins and fish. *Ethology and Sociobiology* 9: 149–79.

Norris, K. S., W. E. Stuntz, and W. Rogers. 1978. The behavior of porpoises and tuna in the eastern tropical Pacific yellowfin tuna fishery: Preliminary studies. Final report to the Marine Mammal Commission. PB-283 970.

Norris, K. S., B. Würsig, R. S. Wells, and M. Würsig, eds. 1994. *The Hawaiian Spinner Dolphin*. Berkeley: University of California Press.

Nousek, A. E., P. J. B. Slater, C. Wang, and P. J. O. Miller. 2006. The influence of social affiliation on individual vocal signatures of northern resident killer whales (*Orcinus orca*). *Biology Letters* 2 (4): 481–84.

Nowacek, D. P., L. H. Thorne, D. W. Johnston, and P. Tyack. 2007. Responses of cetaceans to anthropogenic noise. *Mammal Review* 37: 81–115.

Östman, J. S. O. 1994. Social organization and social behavior of Hawai'ian spinner dolphins (*Stenella longirostris*). Ph.D. diss., University of California, Santa Cruz.

Ottoni, E., and M. Mannu. 2003. Spontaneous use of tools by semifree-ranging Capuchin monkeys. In *Animal Social Complexity: Intelligence, Culture, and Individualized Societies*, ed. F. B. M. de Waal and P. L. Tyack, 440–43. Cambridge, MA: Harvard University Press.

Pabst, D. A., S. A. Rommel, and W. A. McLellan. 1999. The functional morphology of marine mammals. In *Biology of Marine Mammals*, ed. J. E. Reynolds III and S. A. Rommel, 15–72. Washington, DC: Smithsonian Institution Press.

Pack, A. A. 1994. Cross-modal recognition of complexly-shaped objects by a bottlenosed dolphin using vision and echolocation. Ph.D. diss., University of Hawaii.

Pack, A. A., and L. Herman. 2006. Dolphin social cognition and joint visual attention: Our current understanding. *Aquatic Mammals* 32 (4): 443–60.

———. 2007. The dolphin's (*Tursiops truncatus*) understanding of human gazing and pointing: Knowing what and where. *Journal of Comparative Psychology* 121 (1): 34–45.

Packer, C. 1994. *Into Africa*. Chicago: University of Chicago Press.

Patterson, F. G. 1978. The gestures of a gorilla: Language acquisition in another pongid. *Brain and Language* 5: 72–97.

Patterson, I. A. P., R. J. Reid, B. Wilson, K. Grellier, H. M. Ross, and P. M. Thompson. 1998. Evidence for infanticide in bottlenose dolphins: An explanation for violent interactions with harbour porpoises? *Proceedings of the Royal Society of London, B: Biological Sciences* 265: 1–4.

Paulos, R. D., K. M. Dudzinski, and S. A. Kuczaj. 2008. Non-vocal communication in the Atlantic spotted dolphin (*Stenella frontalis*) and the Indo-Pacific bottlenose dolphin (*Tursiops aduncus*). *Journal of Ethology* 26: 153–64.

Payne, K. 1998. *Silent Thunder in the Presence of Elephants*. New York: Simon and Schuster.

Pepperberg, I. M., and M. V. Brezinsky. 1991. Acquisition of a relative class concept by an African gray parrot (*Psittacus erithacus*): Discriminations based on relative size. *Journal of Comparative Psychology* 105 (3): 286–94.

Perrin, W. F. 1970. Color patterns of the Eastern Pacific spotted porpoise *Stenella graffmani* Lönnberg (Cetacea, Delphinidae). *Zoologica* 54: 135–42.

Perrin, W. F., B. Würsig, and J. G. M. Thewissen, eds. 2002. *Encyclopedia of Marine Mammals*. San Diego: Academic.

Peters, R. 1980. *Mammalian Communication*. Belmont, CA: Brooks/Cole.

Petricig, R. O. 1993. Diet patterns of "strand feeding" behavior by bottlenose dolphins in South Carolina salt marshes. Abstracts of the Tenth Biennial Conference on the Biology of Marine Mammals, November 11–15, 1993, Galveston, TX.

———. 1995. Bottlenose dolphins (*Tursiops truncatus*) in Bull Creek, South Carolina. Ph.D. diss., University of Rhode Island.

Philips, M., and S. N. Austad. 1990. Animal communication and social evolution. In *Interpretation and Explanation in the Study of Animal Behavior,* vol. 1, ed. M. Bekoff and D. Jamieson, 254–68. Boulder, CO: Westview.

Pinker, S. 1995. *The Language Instinct.* New York: Harper Perennial.

Pliny the Elder. 1940. *Natural History.* Vol. 3, books 8–11, trans. H. Rackham. Loeb Classical Library. Cambridge, MA: Harvard University Press.

Plotnik, J. M., F. B. M. de Waal, and D. Reiss. 2006. Self-recognition in an Asian elephant. *Proceedings of the National Academy of Sciences of the United States* 103: 17053–57.

Podos, J., V. M. F. Da Silva, and M. R. Rossi-Santos. 2002. Vocalizations of Amazon river dolphins, *Inia geoffrensis:* Insights into evolutionary origins of Delphinidae whistles. *Ethology* 108: 1–12.

Polasek, L. K., and R. W. Davis. 2001. Heterogeneity of myoglobin in the locomotory muscles of five cetacean species. *Journal of Experimental Biology* 204 (2): 209–15.

Poole, J. H. 1996. *Coming of Age with Elephants: A Memoir.* London: Hyperion.

Povinelli, D. J., and S. Giambrone. 2001. Reasoning about beliefs: A human specialization? *Child Development* 72 (3): 691–95.

Povinelli, D. J., G. G. Gallup, Jr., T. J. Eddy, and D. T. Bierschwale. 1997. Chimpanzees recognize themselves in mirrors. *Animal Behaviour* 53 (5): 1083–88.

Premack, D. 1971. Language in chimpanzee? *Science* 172 (985): 808–22.

Pryor, K. W. 1973. Behavior and learning in porpoises and whales. *Naturwissenschaften* 60: 412–20.

———. 1986. Reinforcement training as interspecies communication. In *Dolphin Cognition and Behavior: A Comparative Approach,* ed. R. J. Schusterman, J. A. Thomas, and F. G. Wood, 253–60. Hillsdale, NJ: Lawrence Erlbaum Associates.

———. 1990a. Non-acoustic communication in small cetaceans: Glance, touch, position, gesture, and bubbles. In *Sensory Abilities of Cetaceans: Laboratory and Field Evidence,* ed. J. A. Thomas and R. A. Kastelein, 537–44. New York: Plenum.

———. 1990b. Dolphin-swim behavioral observation program: Suggestions for a research protocol. Attachment C in the final report on the workshop to develop a recommended study design for evaluating the relative risks and benefits of swim-with-the-dolphin programs. Report authors R. S. Wells and S. Montgomery. Unpublished report to the Marine Mammal Commission, Washington, DC.

Pryor, K. W., R. Haag, and J. O'Reilly. 1969. The creative porpoise: Training for novel behavior. *Journal of Experimental Analysis of Behavior* 12 (4): 653–61.

Pryor, K., and I. Kang. 1980. Social behavior and school structure in pelagic porpoises (*Stenella attenuata* and *S. longirostris*) during purse seining for tuna. Final report to the National Marine Fisheries Service LJ-80-11C.

Pryor, K., J. Lindbergh, S. Lindbergh, and R. Milano. 1990. A dolphin-human fishing cooperative in Brazil. *Marine Mammal Science* 6 (1): 77–82.

Pryor, K., and K. S. Norris, eds. 1991. *Dolphin Societies: Discoveries and Puzzles*. Berkeley: University of California Press.

Pryor, K., and I. K. Shallenberger. 1991. Social structure in spotted dolphins (*Stenella attenuata*) in the tuna purse seine fishery in the eastern tropical Pacific. In *Dolphin Societies: Discoveries and Puzzles*, ed. K. Pryor and K. S. Norris, 161–98. Berkeley: University of California Press.

Purton, A. C. 1978. Ethological categories of behavior and some consequences of their conflation. *Animal Behaviour* 26: 653–70.

Reidenberg, J. S., and J. T. Laitman. 1988. Existence of vocal folds in the larynx of Odontoceti (toothed whales). *Anatomical Record* 221: 884–91.

Reiss, D., and L. Marino. 2001. Mirror self-recognition in the bottlenose dolphin: A case of cognitive convergence. *Proceedings of the National Academy of Sciences of the United States* 98 (10): 5937–42.

Rendell, L. E., J. N. Matthews, A. Gill, J. C. D. Gordon, and D. W. MacDonald. 1999. Quantitative analysis of tonal calls from five odontocete species, examining interspecific variation. *Journal of Zoology* (London) 249: 403–10.

Rendell, L., and H. Whitehead. 2001. Culture in whales and dolphins. *Behavioral and Brain Sciences* 24 (2): 309–24.

———. 2003. Vocal clans in sperm whales (*Physeter macrocephalus*). *Proceedings of the Royal Society B: Biological Sciences* 270: 225–31.

Reynolds, J. E., III, and S. A. Rommel. 1999. *Biology of Marine Mammals*. Washington, DC: Smithsonian Institution Press.

Reynolds, J. E., III, R. S. Wells, and S. D. Eide. 2000. *The Bottlenose Dolphin: Biology and Conservation*. Gainesville: University of Florida Press.

Richardson, W. J., Jr., C. R. Greene, C. I. Malme, and D. H. Thomson. 1998. *Marine Mammals and Noise*. San Diego: Academic.

Richardson, W. J., Jr., C. R. Greene, C. I. Malme, D. H. Thomson, and B. Würsig. 1991. Effects of noise on marine mammals. U.S. Dept. of Interior, Contract No. 14-12-0001-30362, Minerals Management Service, Herndon, VA.

Richter, C., S. Dawson, and E. Slooten. 2006. Impacts of commercial whale watching on male sperm whales at Kaikoura, New Zealand. *Marine Mammal Science* 22: 46–63.

Ridgway, S. H. 2000. The auditory central nervous system of dolphins: Hearing by whales and dolphins. In *Handbook of Auditory Research*, ed. W. W. L. Au, A. N. Popper, and R. R. Fay, 273–93. New York: Springer.

Ridgway, S., B. L. Scronce, and J. Kanwisher. 1969. Respiration and deep diving in the bottlenose porpoise. *Science* 166: 1651–54.

Roper, T. J. 1986. Cultural evolution of feeding behaviour in animals. *Science Progress* 70: 571–83.

Rose, N. A., R. H. Farinato, and S. Sherwin, eds. 2006. *The Case against Marine Mammals in Captivity*. Washington, DC: Humane Society of the United States and World Society for the Protection of Animals.

Rossbach, K. A., and D. L. Herzing. 1997. Underwater observations of benthic feeding bottlenose dolphins (*Tursiops truncatus*) near Grand Bahama Island, Bahamas. *Marine Mammal Science* 13 (3): 498–503.

Saayman, G. S., and C. K. Taylor. 1979. The socioecology of humpback dolphins (*Sousa* sp.). In *Behavior of Marine Mammals: Current Perspectives in Research*, vol. 3, ed. H. E. Winn and B. L. Olla, 165–226. New York: Plenum.

Samuels, A., L. Bejder, R. Constantine, and S. Heinrich. 2003. Swimming with wild cetaceans with a special focus on the Southern Hemisphere. In *Marine Mammals: Fisheries, Tourism and Management Issues*, 277–303. Collingwood, Australia: CSIRO.

Samuels, A., L. Bejder, and S. Heinrich. 2000. *A Review of the Literature Pertaining to Swimming with Wild Dolphins*. Silver Spring, MD: Marine Mammal Commission.

Samuels, A., and T. Spradlin. 1995. Quantitative behavioral study of bottlenose dolphins in swim-with-dolphin programs in the United States. *Marine Mammal Science* 11 (4): 510–19.

Santos, M. C. O. 1997. Lone sociable bottlenose dolphin in Brazil: Human fatality and management. *Marine Mammal Science* 13 (2): 355–56.

Sargeant, B. L., J. Mann, P. Berggren, and M. Krützen. 2005. Specialization and development of beach hunting, a rare foraging behavior, by bottlenose dolphins (*Tursiops* sp.). *Canadian Journal of Zoology* 83: 1400–1410.

Savage-Rumbaugh, E. S., R. A. Sevcik, D. M. Rumbaugh, and E. Rubert. 1985. The capacity of animals to acquire language: Do species differences have anything to say to us? *Philosophical Transactions of the Royal Society B: Biological Sciences* 308: 177–85.

Sayigh, L. S., P. L. Tyack, R. S. Wells, and M. D. Scott. 1990. Signature whistles of free-ranging bottlenose dolphins *Tursiops truncatus:* Stability and mother-offspring comparisons. *Behavioral Ecology and Sociobiology* 26: 247–60.

Sayigh, L. S., P. L. Tyack, R. S. Wells, M. D. Scott, and A. B. Irvine. 1995. Sex differences in signature whistle production of free-ranging bottlenose dolphins, *Tursiops truncatus. Behavioral Ecology and Sociobiology* 36: 171–77.

Schotten, M., M. O. Lammers, K. D. Sexton, and W. W. L. Au. 2005. A new underwater portable four-channel acoustic/video recording unit to study echolocation and communication of individual wild dolphins. Abstracts of the Sixteenth Biennial Conference on the Biology of Marine Mammals, December 12–16, 2005, San Diego.

Schusterman, R. J., and R. Gisiner. 1988. Artificial language comprehension in dolphins and sea lions: The essential cognitive skills. *Psychological Record* 38: 311–48.

Senate Select Committee on Animal Welfare. 1985. Dolphins and whales in captivity. Canberra: Australian Government Publishing Service.

Seyfarth, R., and D. Cheney. 1999. Inside the mind of a monkey. In *Readings in Animal Cognition,* ed. M. Bekoff and D. Jamieson, 337–43. Cambridge, MA: MIT Press.

Shane, S. H. 1977. The population biology of the Atlantic bottlenose dolphin, *Tursiops truncatus,* in the Aransas Pass area of Texas. M.A. thesis, Texas A&M University.

———. 1990. Behavior and ecology of the bottlenose dolphin at Sanibel Island, Florida. In *The Bottlenose Dolphin,* ed. S. Leatherwood and R. R. Reeves, 245–65. San Diego: Academic.

Shane, S., R. S. Wells, and B. Würsig. 1986. Ecology, behavior and social organization of the bottlenose dolphin: A review. *Marine Mammal Science* 2: 34–63.

Silber, G. K., and D. Fertl. 1995. Intentional beaching by bottlenose dolphins (*Tursiops truncatus*) in the Colorado river delta, Mexico. *Aquatic Mammals* 21 (3): 183–86.

Simard, P., and S. Gowans. 2004. Two calves in echelon: An alloparental association in Atlantic white-sided dolphins (*Lagenorhynchus acutus*)? *Aquatic Mammals* 30 (2): 330–34.

Similiä, T. 1997. Behavioral ecology of killer whales in northern Norway. Ph.D. diss., Norwegian College of Fisheries Science, University of Tromsø.

Simmonds, M. P. 2006. Into the brains of whales. *Applied Animal Behaviour Science* 100 (1–2): 103–16.

Skinner, B. F. 1938. *The Behavior of Organisms: An Experimental Analysis.* New York: Appleton-Century-Crofts.

Smith, B. 2003. The discovery and development of dolphin-assisted therapy. In *Between Species: Celebrating the Dolphin-Human Bond,* ed. T. Frohoff and B. Peterson, 239–46. San Francisco: Sierra Club Books.

Smith, W. J. 1990. Communication and expectations: A social process and the cognitive operations it depends upon and influences. In *Interpretation and Explanation in the Study of Animal Behavior, vol. 1., ed. M. Bekoff* and D. Jamieson, 234–53. Boulder, CO: Westview.

———. 1991. Animal communication and the study of cognition. In *Cognitive Ethology: The Minds of Other Animals*, ed. C. A. Ristau, 209–30. Hillsdale, NJ: Lawrence Erlbaum Associates.

Smolker, R., A. Andrew, R. Connor, J. Mann, and P. Berggren. 1997. Sponge carrying by dolphins (Delphinidae, *Turiops* sp.): A foraging specialization involving tool use? *Ethology* 103 (6): 454–65.

Smolker, R. A., J. Mann, and B. Smuts. 1993. Use of signature whistles during separations and reunions by wild bottlenose dolphin mothers and infants. *Behavioral Ecology and Sociobiology* 33: 393–402.

Smuts, B. 1988. Primates, dolphins, and the origins of intelligence. *American Scientist* 76: 494–99.

Soto, N. A., M. Johnson, P. T. Madsen, P. L. Tyack, A. Bocconcelli, and J. F. Borsani. 2006. Does intense ship noise disrupt foraging in deep-diving Cuvier's beaked whales (*Ziphius Cavirosris*)? *Marine Mammal Science* 22 (3): 690–99.

Southall, B. 2005. Shipping noise and marine mammals: A forum for science, management and technology. Report from the Symposium, May 18–19, 2004, Arlington, VA. Silver Spring, MD: NOAA Fisheries, Acoustics Program.

Spinka, M., R. C. Newberry, and M. Bekoff. 2001. Mammalian play: Can training for the unexpected be fun? *Quarterly Review of Biology* 76: 1–28.

Steiner, W. W. 1981. Species-specific differences in pure tonal whistle vocalizations of five western North Atlantic dolphin species. *Behavioral Ecology and Sociobiology* 9: 241–46.

Sweeney, J. C. 1990. Marine mammal behavioral diagnostics. In *CRC Handbook of Marine Mammal Medicine: Health, Disease, and Rehabilitation*, ed. L. A. Dierauf, 53–72. Boston: CRC.

Tarpley, R. J., and S. H. Ridgway. 1994. Corpus callosum size in delphinid cetaceans. *Brain Behavior and Evolution* 44: 156–65.

Taylor, C. K., and G. S. Saayman. 1972. The social organisation and behaviour of dolphins (*Tursiops aduncus*) and baboons (*Papio ursinus*): Some comparisons and assessments. *Annals of the Cape Provincial Museums Natural History* 9 (2): 11–49.

Teleki, G. 1972. The omnivorous chimpanzee. *Scientific American* 228: 2–12.

Thewissen, J. G. M. 2002. Hearing. In *Encyclopedia of Marine Mammals*, ed. W. F. Perrin, B. Würsig, and J. G. M. Thewissen, 570–74. San Diego: Academic.

Thewissen, J. G. M., and E. M. Williams. 2002. The early radiations of Cetacea (Mammalia): Evolutionary pattern and developmental correlations. *Annual Review of Ecological Systems* 33: 7390.

Thewissen, J. G. M., M. J. Cohn, L. S. Stevens, S. Bajpai, J. Heyning, and W. E. Horton, Jr. 2006. Developmental basis for hind-limb loss in dolphins and origin of the cetacean body plan. *Proceedings of the National Academy of Sciences of the United States* 103 (22): 8414–18.

Thewissen, J. G. M., E. M. Williams, and S. T. Hussain. 2000. Anthracobunidae and the relationships among Desmostylia, Sirenia and Proboscidea. *Journal of Vertebrate Paleontology* 20: 73A.

Trivers, R. L. 1971. The evolution of reciprocal altruism. *Quarterly Review of Biology* 46: 35–57.

Trone, M., S. Kuczaj, and M. Solangi. 2005. Does participation in dolphin-human interaction programs affect bottlenose dolphin behaviour? *Applied Animal Behaviour Science* 93: 363–74.

Tschudin, A. J. P. C. 2001. "Mindreading" mammals? Attribution of belief tasks with dolphins. *Animal Welfare* 10: S119–27.

———. 2006. Belief attribution tasks with dolphins: What social minds can reveal about animal rationality. In *Rational Animals?* ed. S. Hurley and M. Nudds, 413–38. Oxford: Oxford University Press.

Tschudin, A., J. Call, R. I. Dunbar, G. Harris, and C. van der Elst. 2001. Comprehension of signs by dolphins (*Tursiops truncatus*). *Journal of Comparative Psychology* 115 (1): 110–15.

Twiss, J. R., II, and R. R. Reeves, eds. 1999. *Conservation and Management of Marine Mammals*. Washington, DC: Smithsonian Institution Press.

Tyack, P. 2003. Research program to evaluate effects of manmade noise on marine mammals in the Ligurian Sea. ACCOBAMS. Document CS2/Inf.13.

Tyack, P. L., and L. S. Sayigh. 1997. Vocal learning in cetaceans. In *Social Influences on Vocal Development*, ed. C. T. Snowden and M. Hausberger, 419–57. Cambridge: Cambridge University Press.

Van der Pol, B. A. E., J. G. F. Worst, and P. van Andel. 1995. Macro-anatomical aspects of the cetacean eye and its imaging system. In *Sensory Systems of Aquatic Mammals*, ed. R. A. Kastelein, J. A. Thomas, and P. E. Nachtigall, 409–14. Woerden, Netherlands: De Spil.

Vater, M., and M. Kössl. 2004. The ears of whales and bats. In *Echolocation in Bats and Dolphins*, ed. J. A. Thomas, C. Moss, and M. Vater, 89–99. Chicago: University of Chicago Press.

Von Frisch, K. 1967. *The Dance Language and Orientation of Bees*. Trans. L. Chadwick. Cambridge, MA: Harvard University Press.

Walther, F. R. 1984. *Communication and Expression in Hoofed Animals*. Bloomington: Indiana University Press.

Wang, D., B. Würsig, and W. Evans. 1995a. Comparisons of whistles among seven odontocete species. In *Sensory Systems of Aquatic Mammals*, ed. R. A. Kastelein, J. A. Thomas, and P. E. Nachtigall, 299–323. Woerden, Netherlands: De Spil.

———. 1995b. Whistles of bottlenose dolphins: Comparisons among populations. *Aquatic Mammals* 21: 65–77.

Waples, K. A., and N. J. Gales. 2002. Evaluating and minimising social stress in the care of captive bottlenose dolphins (*Tursiops truncatus*). *Zoo Biology* 21: 5–26.

Warburton, D. M. 1991. Stress and distress in response to change. In *Primate Response to Environmental Change*, ed. H. O. Box, 337–56. New York: Chapman and Hall.

Watkins, W. A., and D. Wartzok. 1985. Sensory biophysics of marine mammals. *Marine Mammal Science* 1 (3): 219–60.

Weilgart, L. S., and H. Whitehead. 1993. Coda communication by sperm whales (*Physeter macrocephalus*) off the Galapagos Islands. *Canadian Journal of Zoology* 71: 744–52.

Weir, A. A., and A. Kacelnik. 2006. A New Caledonian crow (*Corvus moneduloides*) creatively re-designs tools by bending or unbending aluminium strips. *Animal Cognition* 9 (4): 317–34.

Wells, R. S. 1991. Bringing up baby. *Natural History* August: 56–62.

Wells, R. S., A. B. Irvine, and M. D. Scott. 1980. The social ecology of inshore odontocetes. In *Cetacean Behavior: Mechanisms and Functions*, ed. L. M. Herman, 263–317. New York: John Wiley and Sons.

Wells, R. S., M. D. Scott, and A. B. Irvine. 1987. The social structure of free-ranging bottle-nose dolphins. In *Current Mammalogy*, ed. H. H. Genoways, 247–305. New York: Plenum.

Wilke, M., M. Bossley, and M. Doak. 2005. Managing human interactions with solitary dolphins. *Aquatic Mammals* 31 (4): 427–33.

Williams, R., and D. Lusseau. 2006. A killer whale social network is vulnerable to targeted removals. *Biology Letters* 2 (4): 497–550.

Williams, R. M., A. W. Trites, and D. E. Bain. 2002. Behavioural responses of killer whales (*Orcinus orca*) to whale-watching boats: Opportunistic observations and experimental approaches. *Journal of Zoology (London)* 256: 255–70.

Williams, T. M., W. A. Friedl, M. L. Fong, R. M. Yamada, P. Sedivy, and J. E. Haun. 1992. Travel at low energetic cost by swimming and wave-riding bottlenose dolphins. *Nature* 355: 821–23.

Worsham, R. W., and M. R. D'Amato. 1973. Ambient light, white noise, and monkey vocalization as sources of interference in visual short-term memory in monkeys. *Journal of Experimental Psychology* 99: 99–105.

Worthy, G. A. J., and E. F. Edwards. 1990. Morphometric and biochemical factors affecting heat loss in a small temperate cetacean (*Phocoena phocoena*) and small tropical cetacean (*Stenella attenuata*). *Physiological Zoology* 63: 432–42.

Würsig, B., T. R. Kieckhefer, and T. A. Jefferson. 1990. Visual displays for communication in cetaceans. In *Sensory Abilities of Cetaceans*, ed. J. Thomas and R. Kastelein, 545–49. New York: Plenum.

Würsig, B., and M. Würsig. 1979. Behavior and ecology of the bottlenose dolphin, *Tursiops truncatus*, in the South Atlantic. *Fishery Bulletin* 77: 399–412.

———. 1980. Behavior and ecology of the dusky dolphin, *Lagenorhynchus obscurus*, in the South Atlantic. *Fishery Bulletin* 77: 871–90.

Wynne, C. D. L. 2004. *Do Animals Think?* Princeton, NJ: Princeton University Press.

Xitco, M. J., Jr. 1988. Mimicry of modeled behaviors in bottlenose dolphins. M.A. thesis, University of Hawaii.

Xitco, M. J., J. D. Gory, and S. A. Kuczaj. 2001. Spontaneous pointing by bottlenose dolphins (*Tursiops trucatus*). *Animal Cognition* 4: 115–23.

———. 2004. Dolphin pointing is linked to the attentional behavior of a receiver. *Animal Cognition* 7 (4): 231–38.

Xitco, M. J., and H. L. Roitblat. 1996. Object recognition through eavesdropping: Passive echolocation in bottlenose dolphins. *Animal Learning and Behavior* 24 (4): 355–65.

Yurk, H., L. Barrett-Lennard, J. K. B. Ford, and C. O. Matkin. 2002. Cultural transmission within maternal lineages: Vocal clans in resident killer whales in southern Alaska. *Animal Behavior* 63: 1103–19.

Illustration Credits

Unless otherwise indicated below, all black-and-white photographs and color plates are copyright by John Anderson, Terramar Productions.
All illustrations were created by John Norton.

Will Anderson: plate of Toni being greeted by a gray whale
Kathleen M. Dudzinski: pages 30, 33, 52, 60, 61, 68, 71; plate of two young spotted dolphins vertical and upside down
Cathy Kinsmen: page 137; plate of beluga with people in boat reaching down to touch her

Index